THOMPSON'S GUIDE
TO FRESHWATER FISHES

ALSO BY PETER THOMPSON
The Game Fishes of New England
and Southeastern Canada

THOMPSON'S GUIDE TO FRESHWATER FISHES

How to Identify the
Common Freshwater Fishes
of North America

How to Keep Them
in a Home Aquarium

Text and illustrations by
PETER THOMPSON

Houghton Mifflin Company   Boston
1985

Library of Congress Cataloging in Publication Data

Thompson, Peter, date
Thompson's Guide to freshwater fishes.

Bibliography: p.
Includes index.
1. Aquariums.   2. Fishes, Fresh-water—United States.
3. Fishes, Fresh-water—Canada.   4. Fishes, Freshwater—United States—Identification.   5. Fishes,
Fresh-water—Canada—Identification.   6. Fishes—
United States—Identification.   7. Fishes—Canada—
Identification.   I. Title.   II. Title: Guide to
freshwater fishes.

SF457.T448   1985   597.092'973   84-19195
ISBN 0-395-31838-6
ISBN 0-395-37803-6 (pbk.)

Printed in the United States of America

V 10 9 8 7 6 5 4 3 2 1

To Erin, Adrienne, and Dona

Contents

Preface *ix*

Introduction *1*

Catalogue of Fishes

Gars *27*

Bowfins *30*

Freshwater Eels *32*

Herrings *34*

Mooneyes *37*

Trouts *40*

Mudminnows *49*

Pikes *51*

Characins *58*

Minnows *59*

Suckers *98*

Bullhead Catfishes *104*

Pirate Perches *113*

Trout-Perches *114*

Killifishes *116*

Livebearers *120*

Silversides *123*

Sticklebacks *125*

Temperate Basses *131*

Sunfishes *135*

Perches   *164*
Sculpins   *183*

Appendix   *189*
Glossary   *191*
Bibliography   *197*
Index   *199*

Preface

My father was the first to arouse my interest in fish. A thrifty Scotsman, known to everyone as "Red," he taught me the gentle art of angling by example. What I learned from him I seem to have absorbed through the pores of my skin; I don't recall any lectures on the finer points of fly casting or aquatic entomology. It was all an intriguing mystery to me—I could sit for hours poring over the jumbled contents of Dad's tackle box or idly turning the silky smooth handle of a well-used bait-casting reel, my mind filled with images of huge leaping fish.

We had a green wooden rowboat that we kept on the grassy shore of the town water supply, a pond of several hundred acres with a sand and mud bottom. On warm summer afternoons my father would row me and a friend around that seemingly endless body of yellowish water. We would sit quietly trolling earthworms behind silver or gold spinners, anxiously awaiting any strike. We suffered endless, hopeless backlashes whenever we tried our unskilled hands at casting. But any problems were soon forgotten when one of the solid steel or fiberglass rods bent in response to a fish's first run. It was nearly always a White or Yellow Perch or Pumpkinseed that was unsophisticated or hungry enough to accept our haphazardly presented baits. Just catching fish is enough for most kids; but I wanted to keep my fish alive so that I could have a longer look at them. So, whenever I went fishing I carried along a large metal bucket to hold my catch. I would watch, fascinated, as the fish finned quietly or dashed about in the shallow water of the bucket. After keeping them for an hour or so, I would pour them gently back into the pond.

Being able to observe fish in close quarters inspired me to bigger and better fish-keeping projects. I convinced a couple of friends to help me construct a small dam on a brook that flowed through the woods behind our houses. We were all delighted when we found that the fish we introduced to the newly formed pool remained there for at least a couple of weeks before seeking greener pastures. It never occurred to any of us that our stocking program was anything but perfectly legal and in the best interest of both man and nature. One spring day the rains came down in buckets, and the water backing up behind our surprisingly sturdy dam flooded a sizable portion of the nearby golf course. Naturally, golf took precedence over fish-culture, and we watched sadly as an angry greens-keeper rudely dismantled the dam, undoing the stuff of our boyhood dreams.

That setback did little to dampen my interest in fish watching, although a change of tactics was definitely in order. So I started again on a smaller scale. Instead of the pool in the woods, I used a galvanized washtub by my back door. In it I kept whatever aquatic creatures were unfortunate enough to fall into my possession. Only eels seemed capable of escaping from that makeshift aquarium. On several occasions I came upon the withered corpse of a recent escapee as much as 50 or 60 yards from the washtub. After the untimely deaths of a few of these Houdini-types, I gave up on fish and turned my attention to turtles and frogs. These docile amphibians afforded me considerably more peace of mind than the eels had. Even if a frog did jump out, it was entirely capable of surviving out of water.

I came to a watershed in my fish-keeping career when my parents discarded an old soapstone sink in a shady corner of the back yard. I laid claim to the sink and began transforming it into an outdoor aquarium. When I had finished I was convinced that it would make a perfectly adequate home for any self-respecting fish. A few days later I caught a pale hatchery-reared Brook Trout in a nearby river. I can still remember my excitement as I filled a plastic bag with water and pedaled my bicycle home as fast as I could go.

After a short acclimation period the little trout settled nicely into the daily routine of life in my soapstone aquarium. For the next several weeks I spent many happy hours studying the trout as it lay quietly finning in the sun-dappled water. I never actually saw it eat anything, but that was surely not the result of its lacking food. I dropped worms, flies, beetles, millipedes, and assorted grubs into the water, assuring myself that the fish was taking its meals after dark while I was not around to disturb it. There must have been some truth to this assumption, for there were never any leftovers. Just as I was beginning to feel that I had established some sort of rapport with my fish, Fate intervened. One morning I went to the sink and found the trout missing. Wet raccoon paw-prints were the only clue to its whereabouts. It took me some time to recover from the loss of such a wonderful pet, but by then, fish watching had become a permanent part of my life.

I have outgrown the old soapstone sink and worm-baited hooks and moved on to glass aquariums and complex Atlantic salmon flies. But these changes have not lessened the sense of wonder I feel whenever I see a living fish.

Many people have helped with this book, and I would like to give special thanks to Sally Landry, Harry Foster, Julia Fellows, Bill Kenny, Karsten Hartel, and Gennaro Cavacio.

THOMPSON'S GUIDE
TO FRESHWATER FISHES

Introduction

Thompson's Guide to Freshwater Fishes is a basic guide to the common freshwater fishes found in the United States and Canada. It includes those species of fish most likely to be encountered while fishing, snorkeling, or simply peering into the water from a dock or bridge. It also includes a few fishes that are relatively uncommon but that I find unusually attractive or interesting; the selection is ultimately personal, and someone else might well make a different selection. The guide not only tells how to identify native freshwater fishes but how to collect them for an aquarium or fish pond. The first part of the book provides general information on identifying, collecting, and keeping fishes. The second part is an illustrated catalogue covering 113 species of fish in 22 families.

Modern bony fishes have a direct ancestral link to the earth's oldest vertebrates. While other vertebrate groups developed on land, the fishes remained bound to the water. The large number of species that evolved from the early fishes (conservative estimates range from 15,000 to 20,000 species) is some indication of the diverse habitats that fishes occupy. By taking a broad view of the shapes, colors, and patterns into which individual species have evolved, the observer can glimpse the marvelous complexity of which Nature is capable.

Most home aquariums are stocked with brightly colored exotic fishes from tropical and subtropical waters. In recent years, however, an increasing number of aquarists have discovered the advantages of collecting and keeping native fishes. Although many species of native fishes are hardy and easy to care for in captivity, successful collecting requires an understanding of the habits, habitats, and life cycle of each fish. Such an understanding, in turn, depends on the ability to identify the fish you have caught. To identify fish, and to understand them, you need to know, first of all, what makes a fish a fish.

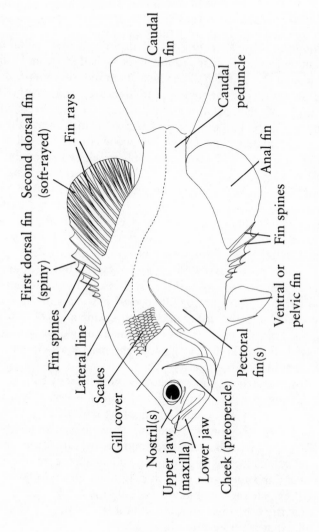

First dorsal fin (spiny)

Second dorsal fin (soft-rayed)

Fin rays

Caudal fin

Fin spines

Caudal peduncle

Lateral line

Anal fin

Scales

Fin spines

Gill cover

Ventral or pelvic fin

Nostril(s)

Pectoral fin(s)

Upper jaw (maxilla)

Lower jaw

Cheek (preopercle)

External Anatomy of a Black Crappie

What Makes a Fish a Fish?

A dictionary will tell you that a fish is a cold-blooded, aquatic vertebrate. Fishes have backbones, streamlined bodies, and round eyes. Four of the most conspicuous external characteristics of fishes are the fins, scales, gills, and lateral line. These characteristics are often important for telling one species—or family—of fish from another, so we will examine them in more detail.

Fins

A fish's fins propel and steer it through the water. A fish has a pair of pectoral fins at its front, and a pair of pelvic fins on its belly. This pairing of fins helps the fish maneuver in a balanced way. Each fish also has an unpaired dorsal fin (or fins) on its top, a single anal fin at its bottom, and a tail fin.

Most fins are thin and flexible, supported by segmented or branched rays. Others are held up by stiff, sharp spines. In some fishes, the bases of the fin spines contain mildly poisonous venom, which may be excreted as a defense against predators.

The most important fin to any fish is the tail or caudal fin. This fin is always soft rayed (although in a few species it may be equipped with a single spine) but may be quite stiff and inflexible. To move forward through the water a fish flexes and relaxes groups of muscles along the rear part of its body; this action drives the caudal fin from side to side, either in wide sweeps or short rapid beats. Often a fish will use its pectoral fins, which roughly correspond to human arms or the wings of a bird, to move ahead or backward. The other fins, which vary in arrangement and shape according to the species, are used as stabilizers or turning devices.

Scales

Modern bony fishes have one of two types of scales. Members of the perch and sunfish families, for instance, have ctenoid scales with tiny teeth on their exposed edges. Other fishes, such as those in the minnow family, have toothless (smooth-edged) scales, known as cycloid scales. In some fishes the scales are difficult to see; eels appear scaleless because their scales are very small, widely spaced, and embedded in the upper layer of the skin. The

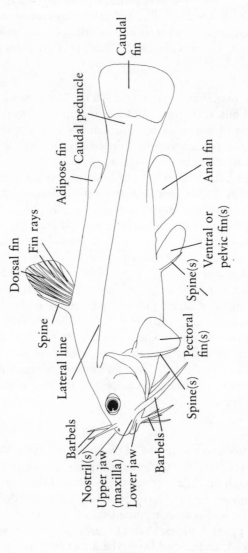

Dorsal fin

Fin rays

Spine

Adipose fin

Caudal peduncle

Caudal fin

Lateral line

Anal fin

Spine(s)

Ventral or pelvic fin(s)

Pectoral fin(s)

Spine(s)

Barbels

Barbels

Nostril(s)

Upper jaw (maxilla)

Lower jaw

External Anatomy of a Black Bullhead

small cycloid scales of the Brook Trout are very smooth and not easy to see at first glance; but a closer inspection will reveal hundreds of tiny scales lying just below the outer layer of the skin.

A layer of mucus, varying in thickness according to species, covers the outer skin of all fishes, making them slippery and difficult to hold. Secreted by glands lying just below the scales, this layer of mucus serves the fish in two very important ways. The mucus is a fish's first defense against invasion by parasites, fungi, and diseases. It also helps the fish move easily through the water by reducing friction.

The outer layers of a fish's skin—the mucus and the scales— also help prevent the animal's body fluids from being diluted by the surrounding water. This outer layer is not impervious, though. It is absorbent and allows water to pass in and out in quantities necessary to maintain the fish's correct chemical balance.

Gills

A gill is a series of flexible, blood-filled filaments attached to thin, crescent-shaped bony structures located at the rear of the fish's head and protected by stiff covers (opercula or gill covers). Gills in fish perform the same function as the lungs in humans. While a lung passes air through tiny blood-rich sacs, a gill passes water over the blood-filled filaments, which extract the oxygen from it. While this oxygen is being taken in, the gills are simultaneously releasing carbon dioxide and toxic wastes back into the water. This entire process takes place in the short time required for a fish to open and close its mouth twice.

Lateral Line

A close look at most species of fishes will reveal a distinct line running roughly along the middle of the fish's side. This is called the lateral line, and it is the external portion of an underlying string of nerve tissue connected directly to the fish's brain. The lateral line consists of a series of pores and tubes that sense the temperature and pressure changes and other disturbances in the surrounding water. It also helps fish move through murky water, to maintain contact with a school, and to navigate at night. A lateral line provides a fish with a very accurate reading of its immediate surroundings. In some fishes the lateral line extends the entire length of the body, including the tail fin, while in others the line reaches only a short distance back from the head. In some species there is no external lateral line. The presence or absence of the lateral line, its length, and the number of scales included in the line is one standard used by ichthyologists and taxonomists to identify fish.

■ IDENTIFYING FISHES

There are two approaches to collecting fish. You can either catch whatever you can and identify it later, or you can seek a specific species in its preferred habitat. In identifying an unfamiliar fish, it is usually easiest to start by figuring out what family it belongs to, and then determine the exact species within that family.

The fishes in this book are arranged by family. Twenty-two families are covered, listed in the table of contents. The family account gives a general description of the characteristics common to most species within that family. The species accounts pinpoint the characteristics you can use to tell one species from others of the same family, or from similar species in different families. Range maps accompanying the text for each species show where that species is found. The maps are only a general guide, since a species may not be found in every watershed within its normal range, and it may be found in an isolated stream or pond well outside its range.

The family and species descriptions are divided into different sections to make it easier to locate specific information. Some of these sections are used only, or primarily, in the family accounts, and some appear only in the species accounts.

Field Marks

These are important identification features that you can use to tell one fish from another. Examples are the size and shape of the body, scales, lateral line, head, mouth, and fins. Is the body round like the Bluegill's or elongated like the Shortnose Gar's? Is the caudal fin deeply forked like the Chain Pickerel's or nearly square like the Fantail Darter's? The field marks that are common to a whole family of fish are described in the family account. The field marks that distinguish one species from others within the same family—or from similar species in different families—are described in the species account.

Adult size: Fish, like most living things, keep growing until they mature; once they reach adult size, the growth process slows down. The length measurements in this book are general guides to the average size of an adult fish, measured from snout to tail fin. In some cases, you may find an adult fish larger or smaller than the range given here. It is often illegal to keep fish under a certain length, so check state and local regulations before you begin collecting.

Color: The color plates at the center of this book show each fish in lateral view. Spend some time studying the plates to familiarize yourself with the colors and patterns of each fish. Color not only enhances a fish's beauty; it is important for identification. What color is your fish? Is it a pale whitish-olive like the Mosquitofish, or a variety of colors, like the Orange-spotted Sunfish?

Habitat: Species of fishes have evolved as part of a specific habitat and are, generally speaking, bound to that type of environment. Large inland bodies of water may contain many different habitats and support fishes with different habitat preferences. For example, a large northern lake might contain salmon or trout, which prefer deep, cold, clear water with a rocky bottom. Largemouth or Smallmouth Bass and members of the pike family thrive in warm shallow water around weeds, stumps, or large boulders. Some minnows prefer warm water and others like cold water. Bottom dwellers such as Bullheads feed and live on muddy bottoms. Small streams or ponds usually support a much less diverse population than a large lake does.

Natural History: Collecting and keeping native fish will give you an increased understanding and appreciation of nature and aquatic life. Once you've identified a fish, you will want to know more about it. This section focuses on migration, the way mating territories are established, proper water temperatures for mating, and spawning rituals. The diets of adults and immatures are also covered.

Local Names: Both common and scientific names of species are given. The first word of the scientific name is the genus and the second word the species. More than one local name is given for many species in this book, since names may vary from region to region. Although scientific names are the only ones recognized by scientists worldwide, common names are usually easier to remember.

Collecting: There are various collecting tools—nets, rod and reel, and traps—and techniques that can be used to collect species, depending on their habits. This section, beginning below, briefly touches upon the proven methods used in collecting the family of fish you are watching.

Handling: Once you have collected a live fish, you'll need to know what water temperature and food it will need in captivity. Some species are much more difficult to keep than others; if a fish I have collected isn't adapting well to aquarium life, I usually return it to the water where I found it. This section gives you tips on how to properly handle and maintain the fishes in your aquarium once you've caught them. Trout, for example, are highly sensitive and need a clean, cool, and highly oxygenated aquarium in order to live. Sunfish, on the other hand, are undemanding in captivity, hardy, and will survive on a variety of foods. See pages 13–23 for general advice on keeping fishes.

■ COLLECTING FISHES

Before setting out on a collecting trip it is very important to check state and local fishing regulations. Some fish are protected and cannot be taken at all. Others have length limits, which may vary from one watershed to another. Different bodies of water often have different rules about how fish can be caught. Open and closed fishing seasons must also be taken into account. If you have an interest in a protected or highly regulated species, it might be possible to obtain a permit from the state department of fish and game to allow the collecting of that species for scientific study.

This book does not include all the fishes that live in the fresh waters of North America but concentrates on those fishes that are relatively well known, abundant (at least locally), and, at some point in their lives, suitable for life in an aquarium. Some of the species I have included are too large for an indoor aquarium but can be kept in outdoor pools or other large impoundments. The species described in this book are by no means the only ones suited for life in captivity. If you want to try collecting other species, use their habits in the wild as a guide to establishing a suitable aquarium.

Equipment

The only essential pieces of fish-collecting equipment are a device for catching the fish and a container in which to transport them. The most commonly used collecting devices are nets and traps. Some other equipment you might use are rod and reel, line, hooks, dip net, minnow trap and bait, nylon or other strong cord, and a bucket or plastic bag.

Knowing the temperature and pH (relative acidity or alkalinity) of the water from which your fish are taken will be a big help in planning the aquarium. With a thermometer and pH meter or litmus paper you can record temperature and pH at the collecting site. A small notebook is useful to record these data and other field observations.

Although many people like nothing better than to slog around barefoot in the muck of a brook or pond, a pair of waterproof boots can help to avoid chills, snakebite, and that old squishy feeling. Hip boots are probably the most useful type; they are less dangerous than chest waders because they keep you from getting into treacherously deep waters. Rubber boots will deteriorate quickly without proper care. Keep them out of sunlight and dry them after each use to help them last as long as possible.

Sunglasses with polarized lenses cut glare from the surface of the water, allowing you to see into any clear body of water.

Nets: The most versatile collecting net is the minnow seine. A seine is made from a rectangular piece of fine-mesh netting with buoyant floats attached to one of the long sides and weights attached to the other. In the water the weighted side is held under the surface and the floats hold the other side up, creating a vertical blockade. The longer and deeper the seine, the more effective it will be at corraling the fish in a given area.

A minnow seine is most easily operated by two people. Each person holds an end of the net, which is supported by poles or cord attached to its corners. The net is drawn slowly through the water as the two people wade along. In ponds or lakes, the net should be drawn toward the shore; in moving water, it should be drawn upstream. When the bottom of the net is lifted to the surface and the sides or ends of the net are drawn together, it creates a kind of bag to hold the fish.

One person can fish a seine by attaching lines to the net's four corners and throwing the net into the water, then pulling it back toward himself. A third method of operating a seine is to set it in a stream, either with poles driven into the bottom or by lines attached to its corners and strung out to bushes on the shore. After setting the net, walk upstream along the shore, then enter the water and wade slowly downstream toward the net. The net must be set so that it can be quickly removed from its place in order to hold the collected fish.

A large, long-handled dip net with a fine mesh bag is a good tool for collecting fish that live in compact schools close to shore. Rainbow Smelt (when they are spawning), Mummichogs, and Brook Silversides can be collected with a dip net. To be most effective, rest the net on the bottom until the fish are swimming over it, then quickly raise it.

A small, fine-mesh dip net is the best tool for moving fish from a traveling container to the aquarium, but it can also double as a collecting tool for very small fish that don't move too fast. Small dip nets can be purchased at any aquarium shop and at many department stores. They are inexpensive, effective, and come in a range of sizes. The best way to use these nets, either at home or in the field, is to move them very slowly so that you disturb the fish as little as possible. It takes considerable concentration and patience to catch fish in this way. The quieter a fish is, the easier it will be to catch. Herding a fish into a corner from which it has no place to escape except into the net works well if the container holding the fish has corners. In buckets it's a matter of catch-as-catch-can.

Another type of net that is good for collecting small fish is a flat, four-cornered net supported by a metal or wooden frame. Lines attached to each corner of the net are tied to a single cord or stout stick. The net is fished by lowering it from a dock or bridge and allowing it to rest on the bottom until fish are swimming over it. It is then quickly raised, trapping the fish in its center.

Minnow Traps: A minnow trap is probably the easiest way to catch collectible specimens. A minnow trap is usually made of galvanized wire mesh and resembles a pair of flower pots connected at their larger ends. There is a mesh funnel-shaped opening at each of the smaller ends. A variety of foods will work well as bait. Probably the most convenient are dry dog food, cat food, and scraps of meat or fish. To prevent these baits from rapidly disintegrating, they should be wrapped in a small fine-mesh bag tied into the trap with a piece of string. The baited trap should be set some distance from the shore of a pond or stream and anchored to a bush or tree with a strong cord. Traps can also be set from a boat in deep water, with a floating marker attached by a cord marking its location. Minnow traps should be allowed to lie undisturbed for several hours. Overnight is best, since many species are most active after dark. A minnow trap is an especially effective way of catching bottom-dwelling fish. It should be checked and emptied at least once every eight hours, since small fish are adversely affected by overcrowding and may injure themselves in trying to escape from a confined space.

Wire minnow traps are inexpensive and readily available from sporting goods or fishing-tackle shops. An alternative is a gallon or half-gallon glass or plastic jar. The mouth of the jar should be 2 to 3 inches wide to make it easier to bait and to remove fish. Bait the jar in the same way you would the wire trap, with the bait bag attached to the anchoring line, which is securely tied around the neck of the jar. The same precautions about setting and checking the wire minnow trap apply to the jar trap as well. Extra caution must be taken when using a glass jar to keep it from breaking.

Angling Equipment: Most fish watchers don't think of angling as a way to collect specimens, and most fishermen don't think of their catch as something that can be kept alive and studied in a home aquarium. But, whether you use a bamboo pole, fly rod, spinning, or casting outfit, angling can be an enjoyable and productive way of collecting fish. Use the lightest tackle and small, barbless hooks to avoid harming the fish you catch.

Transporting Fish

Moving fish from the collecting site to the aquarium is not ordinarily a difficult task. Any problems that might arise usually stem from the special habitat needs of certain species; this is especially true of fish taken from very cold waters. A low water temperature can be maintained by carefully adding ice cubes to the water. Keep in mind that small quantities of water are much more likely to become too warm rather than too cold while you are transporting a fish. An insulated container will greatly reduce the risk of harming the fish, as it will keep the water at close to the temperature that it was in the pond or stream. If the container is uninsulated, be sure to keep it out of direct sunlight. Do not, for instance, leave a bucket of fish in a car parked in the sun. Many novice collectors have been sadly surprised to find how fast sunlight can overheat a small amount of water and kill a fish. A five-gallon plastic pail with a tight-fitting lid is suitable for carrying most fish. Heavy-gauge plastic bags can also be used, although you have to be very careful not to puncture them.

Whatever container you use, it should be filled only ½ to ⅔ full of water, allowing enough space for air, which must be dissolved into the water to provide the fish with the necessary oxygen. As a rule of thumb, each fish should be allowed a minimum of 1 to 2 quarts of water to ensure its getting enough oxygen. Remember that oxygen levels are greatly reduced when the water temperature rises, so special precautions should be taken on hot days to provide more water for each fish. Always avoid overcrowding. Keep in mind that taking fish out of their natural environment and carrying them about in buckets or bags puts them under considerable stress.

Handling Fish

It is important to be careful when handling fish, both for your sake and the fish's. Do not squeeze a fish, as that can easily cause internal injury. Also, you should never handle a live fish with dry hands; this will remove the protective mucus covering the fish. Always remember to wet your hands in water before you touch a fish.

Most native North American fishes are harmless to humans. In fact even the poisonous spines of the catfish family, especially in smaller specimens, do not produce enough venom to inflict anything more than a minor wound. Any fish with spiny fins can cut your hands if you handle it carelessly. To avoid being stuck, simple cradle the fish's body in your palm. If the fish is on a hook, slide your hand down the fish's body from the head, depressing the fins as you go. Usually the heel of the hand should hold the dorsal fins, and the fingers the ventral and anal fins. Members of the perch family will often raise all their fin spines when taken in hand, making them quite difficult to hold; cradling them is the best way to hold these fish. Many members of the perch family have spines on their gill covers that they will flare when handled, but these can be avoided simply by holding the fish around its midsection. Members of the pike family have many sharp teeth, so keep your fingers out of their mouths. Garfish and Bowfins also have sharp teeth.

Catfish have venom sacs at the base of the dorsal and pectoral fin spines. When the fish are taken from the water they will hold these spines stiffly erect, giving you ample opportunity to stick yourself if you aren't careful. The safest way to handle a catfish is by sliding your hand forward along the fish's body, beginning at the tail end and stopping just behind the erected spines.

■ KEEPING FISH

Keeping fish, like collecting them, can be simple or complicated. Your first aquarium may be nothing more than a large bucket or washtub. However, if your aquarium doesn't have a dependable aeration system to provide the fish with oxygen, the water must be changed frequently. Also, an unaerated aquarium should be large—at least 50 gallons—and only a few fish should be kept in it. In most cases, fish should not be kept for more than 30 to 60 days in an unaerated aquarium. Then they should be released into the same body of water from which they were taken. This will prevent new species being accidentally introduced into waters containing pure native fish stocks. Fish will often survive, and even flourish, in waters very different from their preferred habitat. Many fish, especially certain hardy members of the minnow family, can survive under severely adverse conditions.

Fish are more closely bound to water than we are to the air. They take water into their systems to extract oxygen from it and to maintain their bodies' chemical balance. After the fish has used the water, it and the toxic wastes it contains are excreted from the fish's body. This intimacy with the environment makes fish very susceptible to changes in water temperature, acidity, pollution, and a host of other influences.

Water

Aeration and Filtration: The air we breathe usually contains a constant level of oxygen, but the oxygen level in water can fluctuate greatly. The amount of oxygen dissolved in water decreases rapidly as the temperature of the water rises. In running water, oxygen levels are constantly replenished by the water's contact with the air as it breaks up over riffles. To survive in water with reduced oxygen levels, some species of fish regularly move to the surface and take air in through their mouths. Other species seek out cooler water where the oxygen level is higher.

In a permanent aquarium, the problems of oxygen depletion and the buildup of toxic wastes are handled with an air pump and filtration system. The pump is placed outside the tank and feeds air to the filter through a plastic or rubber hose. Most pumps are run by a small electric motor. By using a heavy-duty, large-capacity pump you can run filters for several different tanks from one line.

Filters, like pumps, come in various sizes and shapes. In one type, the filtration medium consists of nylon wool and activated charcoal; this type of filter needs periodic replacement to maintain its efficiency. Another type is the substrate filter, a series of perforated tubes placed in the bottom of the tank and covered with gravel, which circulates the water by a pump. The gravel serves as the filtration medium. Other types of filters are also available.

Consult with an aquarium shop to determine the tank, filter, and pump system that is best suited to your aquarium.

A reliable aeration system will help keep a sufficient supply of dissolved oxygen in the aquarium by bubbling air through the water, as well as moving water to the surface, where it contacts and absorbs oxygen from the air. Because the aerator increases the rate of oxygen exchange, more fish can be kept in an aerated aquarium than in an unaerated aquarium of the same size. Nevertheless, overcrowding must be avoided. Anything smaller than a 10-gallon tank should be used only as a breeding tank or for very small fish. A good rule for stocking an aquarium is to allow 2 to 3 gallons of water for each 3- to 4-inch fish.

Water Temperature: For most of our native species, a heater is not necessary unless the aquarium is in danger of freezing. Keeping the water cool enough can be more of a problem. An aquarium should never be exposed to direct sunlight, or it can easily become too warm for the fish. The water will also tend to become murky looking, as sunlight encourages the growth of algae in the tank.

You should not attempt to keep truly cold-water species, such as the trouts, unless you are quite serious about providing proper aquarium conditions for them. Their preferred temperature range is from 55° to 60°F. This range is also favored by such cold-water species as smelts and sculpins. For these fish you must be especially careful to keep the aquarium in a cool place, or have some sort of cooling unit.

Water Quality: Probably the best type of water with which to fill an aquarium is spring water. If this is inconvenient, tap water is the obvious alternative. Municipal water supplies are almost always chlorinated to some degree, and chlorine in water is dangerous, if not fatal, to fish. Water may be dechlorinated either by allowing it to stand for 2 or 3 days before introducing the fish, or by adding a dechlorinating tablet shortly before the fish are put in.

The relative alkalinity or acidity of water is also important for an aquarium. This is measured on the pH scale, which runs from 0 (very acid) to 14 (very alkaline). The measurement can be made with litmus paper or a pH meter. The pH level of most fresh water is between 6 and 8, which falls in the center of the pH scale. Water that is too acidic or too alkaline can harm fish. To adjust the pH of aquarium water to correspond to the water from which your fish came, you can add sodium bicarbonate ($NaHCO_3$) or some form of limestone to raise the pH and make the water less acidic; or add potassium dihydrogen orthophosphate (Kh_2PO_4) to lower the pH and make the water more acidic. Either of these chemicals should be added to the water in small amounts while you monitor the water's pH with litmus paper or a pH meter. As mentioned before, the litmus paper or meter should also be used to check the pH of the water from which a fish is taken.

Aquariums

Indoor Aquariums: Tanks for an indoor aquarium can be purchased at any pet shop and many department stores. The best tanks are frameless, with glass or Plexiglas walls joined by adhesives. However, for a freshwater aquarium, a tank with a metal or plastic frame may be adequate. If you buy a used tank, or find one in the attic, be sure to check it for leaks before putting any fish in it.

Outdoor Aquariums and Fish Pools: An outdoor aquarium, like any aquarium, should be made of nontoxic materials, such as wood, cement, stone, or porcelain. An outdoor aquarium can be made from a wooden barrel, a discarded bathtub, or a sink; or a traditional ornamental fish pool can be constructed. Plastic should not be used for permanent aquariums, since some of them release small amounts of harmful materials into the water.

In small outdoor aquariums without aerators, the water should be changed frequently to ensure that waste materials do not accumulate to levels harmful to the fish. If the aquarium is light enough, the old water can simply be dumped out while the fish are kept in another container. A more convenient method to remove wastes is to add water through a garden hose until the aquarium is overflowing, and then allow the water to continue to run for 5 or 10 minutes. Heavy-duty aeration systems suitable for outdoor tanks are also available from aquarium or tackle shops.

As with the indoor aquarium, if you put chlorinated water in an outdoor aquarium it must be either aged or dechlorinated before the fish are introduced.

Details on the construction and maintenance of outdoor pools can be found in various books devoted to the subject as well as in periodicals on fish-keeping. You will find suggestions for further reading listed in the Bibliography of this book.

Aquarium Plants: Plants are an integral part of the natural aquatic ecosystem. When the sun is shining they absorb carbon dioxide from the water and release oxygen. This is a great benefit to fish, which utilize oxygen and produce carbon dioxide. Plants also absorb nutrients produced from decaying organic matter, which would otherwise accumulate to harmful levels. But because the average home aquarium is a very tiny body of water, the positive biological effects of plants are reduced to a minimum. They should be regarded primarily as decoration or as shelter for very small fish. If you do decide to grow plants in the aquarium, the best decorating guide is the region from which the fish are collected.

Plants can simply be dug up from the bottom with a trowel or carefully pulled up by hand. Take care to avoid rough handling; keep the roots intact as much as possible and keep the whole plant wet during transport to the aquarium. The plant should be carefully rinsed, with the dead or unhealthy parts removed before it is introduced to the aquarium.

To keep rooted plants healthy, the bottom of the aquarium should be covered with at least 2 to 3 inches of coarse sand or gravel with stones or rocks placed on top. Do not use fine sand, since it will not allow sufficient oxygen to reach the plants' roots.

Because relatively few of our native aquatic plants have been kept in aquariums, some experimentation will be necessary to find the best species from your collecting area. If you are not interested in collecting plants, you can purchase them from an aquarium dealer and, in this way, be assured of having only those plants that have a proven ability to survive in an aquarium. The appendix of native aquatic plants in the back of this book will give you an idea of the various types of plants that might be kept in an aquarium.

Furnishing the Aquarium: Although it is not necessary, any aquarium will be more appealing if it is furnished to resemble the natural habitat of the species it contains. Most warm-water species are associated with aquatic plants, while cold-water species prefer rocky or gravel bottoms with little vegetation. Strictly speaking, each aquarium should be designed for only one type of habitat, and cold-water fishes should be separated from warm-water ones. There is a degree of overlap. Smallmouth Bass, for instance, are equally at home in aquatic vegetation or over bare gravel.

In an aquarium for native freshwater fishes, avoid the brightly colored gravel and plastic or ceramic decorations so often found in tropical-fish aquariums. Try using plants, either rooted or floating, small stumps or branches, and a variety of sizes and shapes of rocks. Any material added to an aquarium, including plants, should be thoroughly rinsed beforehand to remove sediment that would otherwise cloud the water and organisms that might be harmful to the fish. A mild salt and water solution may be used for rinsing and cleaning aquarium materials. Soaps and detergents must not be used, since even small concentrations can be fatal to plants and fish.

Feeding Fishes in the Aquarium

Proper feeding is a very important part of fish-keeping. You can buy dry prepared food or various types of live food from a tropical fish or pet shop. Most native fishes prefer live foods, even if they are not the ones they are used to eating. Live foods available from most aquarium shops include aquatic worms, such as tubifex, and brine shrimp and other aquatic crustaceans, such as daphnia. The immature stages of certain insects (especially midges) are sometimes available. Brine shrimp are generally the most readily available; they can be bought as live adults or live young, as well as freeze-dried. The supply of certain live foods may vary from season to season, making it necessary to alter your fish's diet from time to time. You can avoid shortages in some live foods by raising them in your home. By introducing dried food to your aquarium over an extended period of time, you can train most native fishes to accept it as a steady diet. Members of the minnow family respond especially well to dried food.

A third alternative is to collect fish foods yourself during the warm months. Very fine mesh nets, fished as if they were minnow seines, will scoop up great quantities of plankton and other tiny organisms from a pond or slow-moving stream. These very small foods are excellent for feeding young fishes. Sticking a very fine mesh net into the bottom of a stream and wading downstream toward it, disturbing the bottom as you go, is a good way to gather immature aquatic insects. These will be readily accepted by native fishes. In spring and early summer vast numbers of mosquito larvae can be found in still pools, caught in a small dip net, and fed to your fish. Small grasshoppers and crickets will be hungrily attacked by bass or pickerel. To prevent these live foods from escaping from the aquarium and making your house their house, you should have a tight cover on the tank and feed your fish only as much as they can eat at one time. Collecting your own fish food may not be the most convenient or the most dependable way of getting it, but it can give you a better understanding of the fish you are keeping and their relationship to their environment.

Make sure that the foods you use are suited to the feeding habits of the species you are keeping. For instance, the Blackstripe Topminnow is a surface feeder, preferring to take its food in or slightly below the water's surface film, while the catfishes and suckers are bottom feeders. You can assume that enough food will settle to the bottom of the aquarium to sustain the fishes living there, but you must provide foods that will remain suspended where surface feeders can utilize them. If you feed dry food to surface feeders, it should be in flake form so that it will remain in the surface film for some time before settling to the bottom. Most fishes can feed at a variety of levels, from the surface to the bottom, so you shouldn't encounter much problem if you pay some attention to the feeding habits of those species which feed exclusively at either the top or the bottom. With live foods, you might consider using two types: one that is closely associated with the bottom, such as the various species of aquatic worms, and one that will remain in suspension, such as brine shrimp or insect larvae.

Fish will thrive on small amounts of food. They should be fed only once or twice each day, and they should never be given more food than they can consume in five or ten minutes.

Diseases

Adding new fish to an aquarium always carries the risk of introducing disease. A quarantine tank for new arrivals is the best way to prevent the spread of disease and ensure the health of all your fish. New fish should be kept in quarantine for about two weeks before they are added to any tank containing healthy fish. If during quarantine you observe signs of ill health in any fish, it should be removed and released into the water from which it was collected, or it can be treated with salt, as described below.

A great deal has been written about diseases in exotic aquarium fishes. Some of these data can be applied to our native species. However, since most native fishes have not been long or widely kept in captivity, little information is available on the diseases that affect them. Probably the most common maladies affecting native species are the many types of internal or external parasites that use the fish as either intermediate or final hosts. Fish infested with parasites may appear perfectly healthy for many months before finally succumbing to the infestation. It is therefore difficult to diagnose a fish so infested (especially with internal parasites) until it is too late to save it.

Beyond quarantine, the best means of controlling disease in the aquarium is to keep the aquarium in good working order at all times. Filters should be cleaned regularly and the filtration medium changed; foods should be given to the fish in quantities that can be consumed in a short time. Overcrowding, a major cause of ill-health in fishes, should be carefully avoided.

Should any fish show abnormal behavior, such as erratic swimming motion, unusual inactivity, or being unable to swim upright, it should be removed to a smaller tank that contains water with a small quantity of salt (about ½ teaspoon per gallon) added to it. This is a universal home remedy for ailing fish and seems to work well in many cases. If a fish doesn't respond quickly to the salt treatment, you can release the fish and let nature be the medic, or you may attempt further cures. Since fish are subject to a wide range of illnesses, you may have difficulty in determining the cause of your fish's ill-health. A good place to begin looking for possible solutions is in books on exotic fish (several of which are listed in the Bibliography of this book). Beyond that, you may find information that will help in diagnosing a problem in the scientific literature on individual species.

Breeding Fishes in the Aquarium

Setting up an aquarium in which fish will successfully reproduce is a great challenge, but if you are successful you will be able to observe elaborate mating dances, spawning embraces, unusual nest-building techniques, and parental care of the offspring. The spawning period is the most interesting part of a fish's life, and observation of it is crucial to an understanding of a species' life history.

Our native species are generally much more specific about their spawning habits and requirements than they are about their everyday needs. Unfortunately, many native fish are too large when they reach sexual maturity to be bred in the typical indoor aquarium. However, many, such as the sticklebacks and darters, are of ideal size for the home breeding tank. Procedures for encouraging spawning behavior are described in the individual species accounts under the heading Natural History.

Most fish will not begin mating until they have been given complete privacy. Should a pair of fish spawn in the community tank, the eggs are almost always eaten immediately by other fish. So, the first step in breeding fish is to establish a separate breeding tank, usually one tank for each breeding pair.

Small tanks (such as 5-gallon ones) are especially good for breeding purposes. A tank of this size will take up a minimum of space and can be easily set up, furnished, maintained, and dismantled. The breeding tank must be kept especially clean, and therefore must be provided with a good filtration and aeration system. The tank does not need any furnishings other than those objects or materials required for spawning. Here again, the Natural History sections in the following species accounts describe the specific needs of species.

Once the breeding tank has been established, a pair of fish can be introduced to it. You can either select a pair from the community tank at the approach of the spawning season or plan a collecting trip to coincide with the spawning season of the species you wish to breed. In many species, the onset of the spawning season is announced by changes in color or the growth of tubercles, especially in males. These color changes and growths are the easiest way to select a male and female from each species. Females usually show few, if any, observable changes.

Before putting the pair in the tank, you should carefully inspect each fish to make sure it has no obvious health problems. In some species, such as the sunfishes and sticklebacks, the male should be put into the breeding tank before the female to allow him time to prepare the nest. In other species, the female should be introduced first. (Again, for details, refer to the species accounts.) After both have been put in the breeding tank, watch them closely to see that they do not attack one another. If they do, another pairing must be made.

The fish may need further inducement (beyond simply being together in a small aquarium) to actually begin spawning. In many species, water temperature is the most critical factor in determining the start of reproductive activity. You may need a heater to raise the water temperature enough to trigger spawning. Mating dances and color and fin displays by males, along with increased activity in both fish, are good indications that spawning is about to begin.

You will want to spend as much time as you can near the breeding tank when mating seems imminent, so that you will have a better chance of witnessing the act when it takes place. Some fish, especially in captivity, will attempt to eat all their own eggs immediately after they are fertilized. If that should happen in your breeding tank, simply remove the parents to another aquarium. In the case of the sticklebacks, the male will drive the female from the nest after spawning, at which time she should be removed to insure her safety.

If all has gone well throughout the spawning process, the breeding tank should contain a nest full of healthy eggs and, possibly, their parents. Incubation periods vary greatly among families and species and are heavily influenced by water temperature. The eggs are extremely delicate and can be damaged by changes in their immediate surroundings. During incubation the breeding tank should be left undisturbed and kept at a constant temperature as close as possible to the ideal temperature (see species accounts).

When they have hatched from the egg, the young fish are called fry. The most difficult part of raising very young fish is finding suitable food for them. Their natural food—plankton—is the best food you can offer them. Plankton are tiny aquatic organisms that are usually found suspended in vast numbers in all types of water. During the warm months you can collect your own plankton simply by sweeping a very fine mesh net through the water of a pond or stream. At other times of the year, it will be necessary to purchase very small live foods, such as very young brine shrimp. For most fish, the size of their food should be increased as they grow.

Fry should be kept in the breeding tank until they are large enough to be moved into the community tank without danger of being eaten by larger fish.

CATALOGUE
OF FISHES

■ GARS

Order Lepisosteiformes
Family Lepisosteidae

Field Marks:

· Body is elongated, nearly circular in cross section.
· Body is completely covered with armorlike scales of the ganoid type. The scales are interlocked.
· Lateral line is complete.
· Head is very long, with long, slender, tooth-filled jaws; mouth is very large.
· Seven fins, all soft-rayed: paired pectorals located just behind the lower part of the gill covers; paired ventrals originating at the midpoint of the overall body length; single dorsal and anal fins set far back on the body; and a distinctly rounded caudal.

Habitats: The gars are fishes of warm to temperate waters, generally encountered in large still or slow-moving bodies of water, such as large rivers or lakes.

Natural History: Gars spawn from April to July, earlier in the south and later in the north. The young fish feed on invertebrates, including aquatic insects. The diet of adult gars is made up almost entirely of fish.

Collecting: Gars are all extremely predacious and are quite readily attracted to baited hooks. Their mouths, however, are very bony, making hooking them difficult. Some anglers use small wire snares surrounding the baited hook to catch gars. Gars may also be taken in seines and minnow traps.

Handling: Members of this family are hardy, thriving under a wide range of conditions. They will do well in a properly maintained aquarium. Gars grow quickly, and only small specimens are suitable for the typical home aquarium.

They prefer live foods, especially very small minnows.

Special caution must be taken in handling gars to avoid contact with the many very sharp teeth.

1. Longnose Gar

Lepisosteus osseus

Field Marks:
- Mouth is about ⅔ of the length of the head.
- Jaws are extremely long and slender, filled with many sharp teeth. The snout overhangs the lower jaw.

Adult Size: 2–3 ft. (0.61–0.91 m) average

Color: Adults are usually dark olive-green or olive-brown on the back over creamy or silvery-white sides and belly. Fins are reddish-brown with conspicuous black spots on the caudal and dorsal. The young are dark brown, rust, or black above, with cream or white sides and a rich brown underbelly. Fins are rust to brown with large black spots on the dorsal and caudal. There is a distinct dark dorsal band on the back and a dark lateral band on each side.

Habitats: Adults will be found in deep water while the young are usually associated with dense vegetation in shallow water.

Natural History: Longnose Gars enter shallow, clear streams prior to spawning, which usually takes place from early May through mid-June. During the spawning act, the gravel of the selected shallow riffle area is swept clean by the fish's fins. No nest is prepared, and the eggs are broadcast over the spawning area as they are fertilized. The spawning act is completed in a series of releases of eggs and sperm at irregular intervals. The mating fish, usually several males to each female, engage in much thrashing about during spawning; this has the effect of covering the eggs with the loose bottom gravel. Incubation usually takes 6 to 8 days under good temperature and flow conditions. The fry are equipped with an adhesive disc with which they attach themselves to solid objects while absorbing the yolk sac.

The very young fish feed on aquatic insects and other invertebrates for a very short time. When they have attained a length of several inches, which occurs very rapidly in these fast-growing fish, they feed exclusively on smaller fishes.

Local Names: Billfish, Needlenose Gar, Billy Gar

2. Shortnose Gar

Lepisosteus platostomus

Field Marks:

· Head and snout taper to a point.
· Snout is considerably shorter than in Longnose Gar, slightly over half the total head length.
· From a top view, the snout is broader than in Longnose Gar and overhangs the lower jaw. Jaws are equipped with many very sharp teeth.

Adult Size: About 30 in. (76.2 cm) average

Color: The color is very similar to that of Longnose Gar. Adults are brownish-green to olive-green on the upper back, with cream or white sides and a white underbelly. The young are similar, although colors on the back are often more intense than in adults, with a distinct dark band along the middle of each side. Fins are reddish-brown with distinct black spots on dorsal, caudal, and anal fins.

Habitats: Quiet waters of large rivers and lakes, generally more common in rivers. The young are usually found near vegetation.

Natural History: Shortnose Gars spawn from mid-May through July, depending on location. Spawning fish seek out quiet shallow water with vegetation or other submerged objects on which to lay their eggs. The fertilized eggs are adhesive and attach themselves to plants, roots, or other structure where they remain through about an 8-day incubation period. After hatching, the fry remain near the hatching area for about a week, while they absorb the yolk sac, then move freely about, in search of suitable foods. These young fish spend much of their time drifting close to the water's surface. Like other gars, Shortnose Gars grow very quickly, feeding mainly on aquatic insects, fishes, and crayfish.

Local Names: Short-bill Gar, Billy Gar, Stub-nose Gar

■ BOWFINS

Order Amiiformes
Family Amiidae

Field Marks:

- Body is stout, tapering slightly from head to tail. Body is oval in cross section near the head, more laterally compressed toward the tail.
- Body is covered with large soft scales. Head is not scaled.
- Lateral line is complete.
- Head is large, broad between the eyes.
- Mouth is large, with teeth varying from strong canines in front to much shorter, more rounded ones in back. A gular plate (a thin bony structure situated between the lower jaws) is a distinguishing characteristic of the Bowfin. Upper jaw extends beyond the rear margin of the eye.
- Seven fins: small paired pectorals originating just behind the lower parts of the gill covers; paired ventrals originating just ahead of the midpoint of the body; a short-based, rounded anal; a rounded lobe-like caudal; and a single very long dorsal. All fins are soft-rayed.

Habitats: The Bowfin is usually found in quiet, shallow water near dense vegetation; avoids swift currents or excess turbidity.

Natural History: The Bowfin is the only living representative of this ancient family of fishes. Fossilized remains of bowfins dating from 180 million years ago have been discovered in Europe.

Bowfins spawn in spring, beginning as early as April and continuing into early June. The males enter shallow weedy water, before the females arrive. Each male constructs a nest by removing plants with his mouth and fanning the silt from the clearing with his fins. The female enters the completed nest and releases the eggs, which become adhesive on fertilization. The male lies close beside her and fertilizes the eggs as soon as they are released. The male remains on the nest to guard both the eggs during their 8–10-day incubation period and the fry for a short time after they have hatched. The fry are equipped with adhesive discs on their snouts with which they attach themselves to the nest while absorbing their yolk sacs.

Until they are about 4 in. (10.16 cm) long, young bowfins feed on tiny aquatic crustaceans and other invertebrates. After this they feed almost exclusively on fish.

Like members of the gar family, the bowfin is equipped with a swim bladder that is connected directly to the mouth opening. Both gars and the bowfins are often seen rising to the water's surface to gulp air, which supplements the supply of oxygen taken in through the gills.

Collecting: Small bowfins are easily collected with a minnow seine fished in quiet shallow water or a minnow trap set in the same areas. Larger specimens can be taken by standard angling methods using barbless hooks with live baits.

Handling: Bowfins are hardy and require no special handling in the aquarium. They can be fed a variety of live or dried foods. It should not be kept in a tank with fishes that are much smaller than itself since it may eat them.

3. Bowfin

Amia calva

Field Marks:
· See family Field Marks.

Adult Size: 15–25 in. (38.1–63.5 cm) average

Color: The ground color is yellow-ochre to yellow-olive with darker olive or olive-brown marbled pattern on the back and sides. The belly is cream or white. Yellow-ochre to yellow-brown head with darker olive to brown vertical bars. Dorsal and caudal fins are olive with darker vertical bars. A dark spot, less prominent in females, on the caudal fin is haloed in yellow or orange in males only. Pectoral, ventral, and anal fins are green (with orange tips in males only).

Young specimens are generally lighter overall with the same markings.

Habitats: See family Habitats.

Natural History: See family Natural History.

Local Names: Mudfish, Dogfish, Blackfish, Grindle, Speckled Cat

■ FRESHWATER EELS

Order Anguilliformes
Family Anguillidae

Field Marks:
- Body is extremely elongated, snakelike. The American Eel, the only North American representative of its family, can be confused only with members of the lamprey family from which it can be distinguished by a comparison of the mouths: the eels' upper and lower jaws and teeth resemble those of typical bony fishes while the lampreys' mouths are jawless, disc-like sucking organs that contain a series of rasp-like teeth.
- Scales are very small, widely separated, and embedded in the skin, making the fish appear scaleless.
- Lateral line is complete.
- Head is relatively small with a pointed snout and small eyes. No scales on the head.
- Mouth is large with upper jaw ending behind the rear edge of the eye. Teeth are small but plentiful.
- Eels have one gill slit at the rear of each side of the head.
- Three fins: small paired pectorals originating just behind the gill slits; one continuous fin, beginning at ⅓ of the body length behind the snout and ending just behind the anus, is a fusion of the dorsal, caudal, and anal fins. Ventral fins are absent.

Habitats: The American Eel may be found in all types of water within its range. It is highly tolerant of adverse conditions such as siltation, turbidity, and reduced oxygen levels.

Natural History: The American Eel has a most remarkable natural history that begins, as it does for the European eel, in the warm waters of the Sargasso Sea southwest of Bermuda. Although the eel's spawning act has not been oberved, it is known that all eels make the long journey from their freshwater feeding grounds to their spawning grounds in the Sargasso Sea. At birth, eels are in a larval stage, or leptocephalus, when they look like tiny transparent willow leaves. In this vulnerable state they migrate across vast tracts of open ocean. Not until they reach the estuaries of freshwater rivers and streams do they finally assume their adult shape and color. At this point the males remain in salt and brackish water while the females ascend into fresh water, often traveling hundreds of miles from their point of entry. They may travel through underground waterways and occasionally slither over wet ground, allowing them access to waters that have no apparent connection with the ocean.

Eels are most active after dark, when they do most of their feeding. Their diet includes fishes, mollusks, crustaceans, and aquatic insects. They will also consume decaying animal matter.

Collecting: The best place to find small eels is in the lower reaches of rivers or streams opening directly onto the ocean. They are most plentiful in the spring and early summer when they begin to move inland. Using the glass minnow trap described on page 10 is the easiest way to collect small eels. They can also be taken in small, very fine-mesh, dip nets.

Handling: Young eels can be kept in a community tank. They make no special demands regarding food or water conditions. As they grow they may become aggressive to any other fishes so, they are best kept in large (25–50-gallon) tanks.

4. American Eel

Anguilla rostrata

Field Marks:
· See family Field Marks.

Adult Size: 15–30 in. (38.1–88.9 cm) average

Color: In brackish and salt water eels are gray, black, or olive above, with silvery-white below the lateral line. The fins match the body color, with the dorsal being dark and the anal light. In fresh water the color may be olive, brown, gray, or black above the lateral line and yellow, pale olive, or very light tan below the lateral line. The fins are light olive or light brown, darker on the dorsal and lighter on the ventral.

Habitats: See family Habitats.

Natural History: See family Natural History.

Local Names: Common Eel, Atlantic Eel, Silver Eel, Yellow Eel, Easgann

■ HERRINGS

Order Clupeiformes
Family Clupeidae

Field Marks:
· Body is strongly compressed laterally, with a sharp sawlike edge along the belly.
· Scales are very large and conspicuous.
· Lateral line is absent.
· Head, triangular in profile, has no scales.
· Eyes are large, with transparent eyelids.
· Mouth is small, generally terminal, with either small teeth or no teeth.
· Seven soft-rayed fins: paired pectorals set close behind gill cov-

ers, well down the sides; paired ventrals originating at about the midpoint of the belly; a single long-based anal fin; a broad, deeply forked caudal; and a single triangular dorsal, which may have a threadlike trailing edge.

Habitats: Most herrings spend their lives in salt water. Some, like the American Shad, are anadromous, and a few, including the Gizzard Shad, live only in fresh water. They are most often found in large lakes or rivers, where they travel (in typical herring fashion) in dense schools.

Natural History: Saltwater and anadromous herrings are generally spring spawners. Freshwater species may begin spawning in the spring but often continue throughout the summer.

Herrings feed on very small organisms throughout their lives. Gizzard and Threadfin Shad consume large quantities of algae.

Collecting: Using either minnow seines or large dip nets is the best way to collect freshwater herrings.

Handling: In large aquariums (100-gallon minimum) members of this family may be kept in a small school, which most closely resembles their natural habits. Two or three small specimens may be kept in smaller tanks; however, these fishes should never be crowded in the aquarium. Very small live foods, such as young brine shrimp, are best for members of this family. Fine-grained dry foods may also be given.

5. Gizzard Shad

Dorosoma cepedianum

Field Marks:
- Adult specimens have no teeth.
- Mouth is subterminal, with snout projecting over lower jaw.
- Scales extend onto the base of caudal fin.

• Last ray of dorsal fin is elongated into a thin threadlike projection.

Adult Size: 9–14 in. (22.86–35.56 cm) average

Color: Upper sides and back are deep slate gray or dark blue-green. Remaining body surface is silvery-white with yellow or violet reflective surfaces. Dorsal and caudal fins are gray or gray-brown. Anal, ventrals, and pectorals may be translucent white or flushed with pale yellow.

Habitats: Most often encountered in large schools of varying density in the open water of large lakes, ponds, and slow-moving rivers. May enter brackish water.

Natural History: The Gizzard Shad spawning season begins as early as mid-March and gets progressively later as one moves farther north. At the commencement of spawning, the ripe fish gather into dense groups and swim close to the water's surface, usually in water 3–10 ft. (1–3 m) deep. The mating group begins to roll and tumble about while the eggs are released and fertilized. The fertilized eggs are adhesive and stick to whatever they touch as they sink. Spawning usually takes place at a water temperature of 60°–70°F (16°–21°C).

Incubation of the eggs requires about 4 days under normal conditions. About 5 days after they hatch, the young begin actively feeding. Plankton, especially water fleas, makes up the entire diet of the young fish. When Gizzard Shad are about 8 in. (20 cm) long, their gill rakers increase dramatically in number and length; the gizzard becomes more fully developed, and the gut becomes greatly elongated and convoluted. These changes in their internal organs are in preparation for a shift in diet away from plankton to (almost entirely) algae.

Local Names: Shad, Sawbelly, Mud Shad, Lake Shad, Hickory Shad

6. Threadfin Shad

Dorosoma petenense

Field Marks:
· Similar to Gizzard Shad, except mouth is terminal, with lower jaw projecting slightly beyond snout.

Adult Size: 4–5 in. (10.16–12.7 cm) average

Color: The color of the Threadfin Shad is almost identical to that of the Gizzard Shad. The two species differ in that the Threadfin Shad's fins (except the dorsal) are flushed with bright yellow.

Habitats: The Threadfin Shad is found in habitats similar to those of the Gizzard Shad except that the Threadfin Shad is more often found in water with slight to moderate current.

Natural History: Here again, the Threadfin Shad is nearly identical to the Gizzard Shad. Spawning commences when the water has warmed to about 70°F and may continue throughout the summer months. Unlike the Gizzard Shad, the Threadfin Shad does not change its diet as it matures but continues to feed on a wide range of very small aquatic animals and plants.

■ MOONEYES

Order Osteoglossiformes
Family Hiodontidae

Field Marks:
· Slablike body is strongly compressed laterally.
· Scales are medium-sized, prominent.
· Lateral line is complete, nearly straight from head to tail.
· Head is oval or triangular in profile. Eyes are very large.
· Mouth is medium-sized, terminal, with upper jaw not extending beyond the rear of the eye. Prominent teeth on the jaws.
· Keel partially along midline of the belly.

- Seven soft-rayed fins: paired pectorals set close behind the rear of the gill covers and low on the body; paired ventrals, originating at or slightly ahead of the midpoint of the body; long-based anal set far back; deeply forked caudal with triangular lobes; and triangular dorsal, originating about ⅔ of the total length back from the snout.

Habitats: Mooneyes are fishes of large, slow rivers, small lakes and ponds, and the slow-moving streams that connect them.

Natural History: The Mooneye and Goldeye, found only in North America, are the only surviving species from this ancient family of fishes.

Collecting: Both the Mooneye and the Goldeye are easy to collect, especially as adults. Usually quite plentiful where they occur, they readily accept small baits or artificial flies. Young specimens may be taken in seines or large dip nets.

Handling: Mooneyes are best kept in aquariums of 100 gallons or larger. They grow very quickly and will usually be too large for the average home aquarium. Collecting only small specimens is advisable. Both young and adults are highly suitable for keeping in outdoor pools.

Both species can be fed small live foods and can be conditioned to accept dried foods as well.

7. Mooneye

Hiodon tergisus

Field Marks:

- Rounded, blunt snout very slightly overhangs tip of the lower jaw.
- Keel on the midline of the belly extends only from the anus forward to a point between the ventral fins.
- Anal fin is rounded along the curve of its leading edge, more pointed than in the Goldeye.

Adult Size: 9–11 in. (22.86–27.94 cm) average

Color: Upper sides and back are light olive or light brown. Lower sides are silvery-white with lavender or blue reflections. Belly is white or silvery. Dorsal and caudal fins are light brown. Remaining fins are translucent whitish. All fins are unmarked.

Habitats: The Mooneye is most often found in the shallow waters of clear unsilted lakes, ponds, and rivers.

Natural History: This is a long-lived fish, with males attaining sexual maturity at 3 years of age and females at 5 years. Mooneyes are quite prolific—large females may carry as many as 20,000 eggs. Spawning takes place in April and May after dense schools of mature fish have migrated into tributary rivers and streams. The spawning act has not been documented.

After the eggs have hatched, the young fish pass through a short larval stage, then assume their adult shape. They grow very rapidly, attaining a length of 4–6 in. (10.16–15.24 cm) in their first summer.

The Mooneye's diet consists mainly of immature and adult aquatic and terrestrial insects, usually taken at or just below the water's surface.

Local Names: Toothed Herring, River Whitefish, Freshwater Herring

8. Goldeye
Hiodon alosoides

Field Marks:
· Snout is more pointed than in the Mooneye.
· Eye is a distinct golden yellow.
· Keel along the midline of the belly extends from the anus forward to a point just between the pectoral fins.

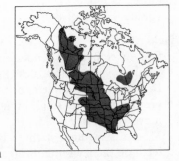

· Anal fin is more square in profile than in the Mooneye.

Adult Size: 14–16 in. (35.56–40.64 cm) average

Color: Back and upper sides are steel blue or gray. Lower sides and belly are silvery-white. Each scale is surrounded by light brown pigment, making the body look tan overall. Fins are uniformly gray. Fins are unmarked.

Habitats: Most frequently encountered in the quiet backwaters of silty, sometimes turbid, rivers and small lakes or ponds that these connect.

Natural History: The Goldeye spawns in the early spring in silty or turbid waters of rivers, streams, or ponds. Spawning has not been observed, but presumably takes place after dark. The semi-bouyant eggs remain suspended a short distance above the bottom during their 2-week incubation period. The fry grow slowly at first but, by September or October of their first year, may attain a length of 3–5 in. (7.62–12.70 cm). The growth rate then slows as the fish reach maturity.

Goldeyes consume a wide variety of foods, although their diet consists mainly of insects during the summer months. Small fishes, mollusks, and crustaceans, and even small mammals, such as mice and shrews, have been found in their stomachs.

■ TROUTS

Order Salmoniformes
Family Salmonidae

Field Marks:
- Body is elongated, slender, in all species.
- Small to very small scales cover the body. Head is not scaled.
- Lateral line is complete, distinct.
- Head is relatively small in most species, generally triangular or narrow oval in profile.
- Mouth is medium-sized to large, with strong, well-developed teeth on the jaws and inside mouth. Upper jaw ends behind the rear edge of the eye.
- Eight soft-rayed fins: paired pectorals originating just behind

the rear edge of the gill covers, set low on the body; paired ventrals originating at about the midpoint of the belly, usually directly below the dorsal fin; a single, short anal fin set back on the belly; a broad caudal fin which may be squarish, rounded, or slightly forked; a single, fleshy adipose fin originating on the mid-line of the back, just ahead of the caudal fin; and a single short-based dorsal near midpoint of the total length.

Habitats: Trouts are all fishes of clear cool waters. They are found in rivers and streams with moderate to swift flows and in cool well-oxygenated ponds, lakes, and bogs. Those populations having access to salt water will often become anadromous.

Natural History: Unless their movement is barred by impassable obstructions, most members of this family move either up or downstream to spawn. Lake- and pond-dwellers enter tributary steams or rivers while stream or river populations generally move upstream or into small tributaries in search of gravel-bottomed riffles, which are this family's preferred spawning sites. The Lake Trout is a notable exception, usually spawning on gravel bars in shallower parts of the lakes and ponds that it inhabits.

Some species spawn in the fall and winter, while others are spring spawners. Prior to spawning, the female excavates a shallow dish-shaped nest in the loose gravel of the stream bottom. These nests are often referred to as redds. The mating pair deposits the fertilized eggs into the nest. The female then turns onto her side and covers the nest by beating her caudal fin on the gravel bottom a short distance upstream. Spawning continues as the female moves upstream a few feet or yards and excavates another nest. This process continues until the eggs are all deposited. The fertilized eggs incubate in the oxygen-rich water percolating through the gravel. No parental care is given the eggs or fry. The fry begin to feed on plankton as soon as the yolk sac is absorbed. The diet of adult trouts consists mainly of aquatic insects and small fishes.

Collecting: Wherever they are found, the trouts are protected by state and federal laws. You may need a special permit to fish for them with a net. A state or provincial fishing license is required to fish for trout with rod and reel.

The several species of trouts described in this book may be caught in a minnow seine or a minnow trap and placed in a stream or brook where trout are known to live. Earthworms contained in a nylon stocking are good bait for trout. To catch trout by angling, use very small live baits or very small artificial flies or lures on barbless hooks. Again, check state regulations on the taking of these fish.

Handling: Trout are especially sensitive to adverse conditions in their environment. An aquarium intended for keeping trout must be kept clean, cool, and highly oxygenated. A substrate filter attached to a powerful pump is the best way to prevent the build-up of toxic wastes. Trout will usually react to water containing unacceptable levels of toxic wastes, usually in the form of nitrates, by thrashing wildly in the tank. Since trout are not usually found in weedy areas in the wild, plants are unnecessary in a tank for trout.

Live foods are best for trout. In summer months you may wish to gather your own live foods from a nearby stream; immature aquatic insects are one of the trouts' favorite natural foods. Urban fish-keepers will find brine shrimp a good food for trout. Trout can also be conditioned to accept dried foods.

9. Golden Trout

Salmo aguabonita

Field Marks:
· Head is small, oval in profile, similar to that of the Rainbow Trout.
· Scales are very small, making the body appear scaleless.
· Mouth is medium-sized. Upper jaw ends below the rear edge of the eyes.
· Caudal fin is slightly forked in both young and adults.

Adult Size: 6–12 in. (15.24–30.48 cm) average

Color: Color varies widely among various strains of this species. Generally the upper back is dark olive-green and the sides are bright orange or yellow. In juvenile trout (and salmon), the broad pink lateral band along the sides and head is marked with a series of oval purplish spots, or parr marks. There is a sprinkling of black spots on the back and sides, concentrated toward the tail. The belly is pink or orange. Pectoral, ventral, and anal fins are pink or yellow, usually matching the color of the belly. Ventral and anal fins have white tips with a black inner border. Dorsal, adipose, and caudal fins are light olive-green or light brown, usually matching the color of the back, and are sprinkled with black spots. The dorsal is tipped with white with a black inner border.

Habitats: The Golden Trout, native only to the headwaters of the Kern River system in the Sierra Nevada, in California, has been successfully introduced into suitable high altitude lakes and streams in the Sierra and Rocky Mountains. The Golden Trout thrives only in clear cold water.

Natural History: Golden Trout that live in running water move upstream in late spring and early summer to spawn. The spawning act is typical of the trouts, with the female digging a series of nests as the mating pair moves progressively upstream. Those populations that have no access to running water use shallow bars and reefs as nesting areas.

Golden Trout depend almost exclusively on aquatic insects for food. The majority of insects available to trout living at high altitudes are caddis flies and midges.

Special Handling: Golden Trout must be kept cool in transport from the collecting site. Ice, the best way of insuring a constant low water temperature, should be added to the water containing the fish in small amounts, so that it does not severely chill the water.

Live brine shrimp, immature native aquatic insects, or dried food (either freeze dried or prepared) are good foods for this species.

10. Cutthroat Trout
Salmo clarki

Field Marks:
- Caudal fin is slightly to moderately forked in both juveniles and adults.
- A bright red or orange streak is visible along lower edge of each gill cover.

Adult Size: 6–13 in. (15.24–33.02 cm) average

Color: Cutthroat Trout vary widely in color and pattern; the bright orange or red streaks, from which this species takes its name, is the most distinguishing characteristic. Overall, the color of the Cutthroat Trout is similar to that of the Rainbow Trout, but lacks the pink lateral band. Young Cutthroat Trout have 7–10 bluish or purplish parr marks along their sides.

Habitats: Found in the clear, cool waters of high altitude ponds and streams. Some populations are anadromous.

Natural History: For decades the Cutthroat Trout has presented scientists with identification problems. Sea-run populations show rather distinct differences from land-locked populations, but the Cutthroat Trout freely interbreeds with both Rainbow and Golden Trout, creating fertile offspring resembling both parents.

Spawning behavior is typical of the trout family. Its diet consists of aquatic insects and small fishes.

Local Names: Coastal Cutthroat, Short-tailed Trout, Yellowstone Trout, Red-throated Trout

11. Rainbow Trout

Salmo gairdneri

Field Marks:
· Body is more strongly laterally compressed than in other trouts.
· Head is slightly shorter and more rounded in profile than in the Brown or Brook Trout.
· Caudal fin is slightly indented in both juveniles and adults.

Adult Size: 6–15 in. (15.24–38.1 cm) average

Color: The back is light brown or olive-green above a pink, red, or silvery lateral band. Lower sides and belly are silvery to white. Young fish have 5–10 bluish or purplish parr marks along their sides, above which are 5–10 dark spots arranged to fall between the parr marks. Body and head are liberally sprinkled with small black, irregularly shaped spots. Dorsal, adipose, and caudal fins are light brown or olive. Pectoral, ventral, and anal fins are tan to light brown and may be tinted with red or yellow. Often a thin white line runs along the leading edge of these fins.

Habitats: Usually found in highly oxygenated areas of streams and rivers with moderate to swift currents. Also occurs in deep, clear lakes and ponds. Many populations are anadromous; these fish are commonly known as Steelhead.

Natural History: Rainbow Trout are late winter and early spring spawners. The female opens and closes the nests before and after each release of eggs and sperm. No parental care is given the eggs or young. After hatching, the fry remain in the gravel bottom before venturing into open water. During this period their food comes from the yolk sac attached to their bellies. After the yolk sac has been consumed, young Rainbow Trout begin feeding on plankton. As they mature, they begin to feed on aquatic and terrestrial insects. Very large adults feed more regularly on fishes.

Local Names: Silver Trout, Kamloops Trout, Steelhead

12. Brown Trout
Salmo trutta

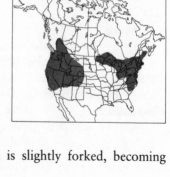

Field Marks:
- Head, especially in adult males, is somewhat longer and more pointed in profile than in the Cutthroat, Rainbow, or Golden Trout.
- Upper jaw extends well beyond rear edge of the eye.
- In young specimens, caudal fin is slightly forked, becoming squared or rounded in adults.

Adult Size: 6–16 in. (15.24–40.64 cm) average

Color: The sides are pale yellow or light tan, the back olive or brown. Belly is creamy-yellow to white. Young Brown Trout have from 9 to 14 blue or violet oval parr marks along their sides. Sides and head are spotted (to widely varying degrees) with black. Often red spots are found along the sides as well (usually close to lateral line). Some of both the red and the black spots are haloed with white or cream color. Pectoral, ventral, and anal fins are pale yellow or tan with white (and sometimes black) stripes along their leading edges. Caudal, adipose, and dorsal fins are olive or brown with black spots. Rear edge of the adipose fin is usually orange or red.

Habitats: Usually prefers those parts of rivers and streams with moderate to slow-moving current. Also thrives in large (usually deep) lakes and ponds. Many populations are anadromous.

Natural History: Strictly speaking, the Brown Trout is not native to North America but was introduced to New England from Europe in 1883. Since then it has been artificially introduced into many waters across this continent. The Brown Trout is more tolerant of adverse conditions than any of our native trouts and has replaced them in many places.

Like other trouts, Brown Trout spawn in the clear headwaters of large rivers and streams or in the tributaries of lakes and ponds. Occasionally it will spawn over shallow reefs and shoals in lakes and ponds. Brown Trout spawn in October and November when the water temperature is between 44° and 48°F (6°–9°C). A 5- or 6-year-old fish will usually produce an average of 2000 eggs. Spawning behavior is typical of the trout family. After absorbing the yolk sac while remaining in the gravel of the stream bed, the fry move into open water and begin feeding on plankton. After this they begin to feed on aquatic and terrestrial insects, mollusks, and crustaceans. Recent studies have shown that Brown Trout are especially fond of the various species of mayflies. The adults feed on a combination of insects and fish. Brown Trout of 15 in. (38.19 cm) and over are often cannibalistic; as a result, many rivers and ponds will have relatively few but very large Brown Trout. Brown Trout are primarily nocturnal feeders, becoming most active on dark, moonless nights.

Local Names: German Brown, Brownie, Salter, Sea Trout, Breac

13. Brook Trout

Salvelinus fontinalis

Field Marks:
· Head is relatively larger than Golden, Cutthroat, and Rainbow Trout; similar in size and shape to that of the Brown Trout.
· Mouth is relatively large, with the upper jaw extending beyond rear edge of the eyes.
· Scales are very small, giving the fish the appearance of being scaleless.
· Caudal fin slightly forked in juveniles and almost perfectly squared along its rear edge in adults.

Adult Size: 6–12 in. (15.24–30.48 cm) average

Color: Upper back is olive-green or olive-brown with many lighter (ochre or pale olive) vermiculations, or wavy lines. Sides range from silvery to yellow, orange, light purple, or light blue-green. The young have 6–9 blueish or purplish oval parr marks, which may remain visible until the fish is quite old, especially in populations living in small cold streams. Belly is pale yellow or orange with black or grayish streaks above a white underbelly. Body is sprinkled with yellow or yellow-ochre spots with a few red spots haloed with blue along the lateral line. Caudal, dorsal, and adipose fins are a reddish-brown and are marked with dark wavy lines. Pectoral, ventral, and anal fins are orange or yellow with white leading edges ahead of thin black lines.

Habitats: Usually encountered in areas of moderate to slow flow, often spring-fed ponds or bogs.

Natural History: Brook Trout spawn in fall, moving upstream or into tributaries in August and September and spawning in September and October. Brook Trout fry hatch in the spring following spawning. Once the yolk sac is absorbed, the young feed on plankton, followed by aquatic and terrestrial insects, mollusks, and crustaceans. Adults will supplement their insect diet with small fish, especially sculpins and minnows, whenever these are available. Brook Trout grow quite slowly, especially in the northern part of their range.

Local Names: Brookie, Squaretail, Speckled Trout, Aurora Trout, Speckled Char, Breac

■ MUDMINNOWS

Order Salmoniformes
Family Umbridae

Field Marks:
· Body is robust and moderately compressed laterally, covered with scales, which are small to large-sized depending on the species. Caudal peduncle is quite deep from top to bottom.
· Lateral line is absent.
· Head is broad, oval with a rounded snout and medium-sized eye.
· Mouth is medium-sized, terminal. Upper jaw ends at about midpoint of the eye. Cheeks and gill covers are scaled.
· Seven soft-rayed fins: paired pectorals originating low on the body just behind the gill covers; paired ventrals originating at about midpoint of the body; a rounded anal fin originating just behind the anus, quite close to the ventrals; a fan-shaped caudal fin; and a short-based fan-shaped dorsal originating shortly behind the midpoint of back.

Habitats: The mudminnows are found in small muddy ponds and slow-moving, weed-filled streams. May also occur in small pools of stagnant water.

Natural History: Mudminnows migrate up small streams in early spring to begin their annual spawning. Spawning may continue over a prolonged period of time, lasting well into summer.

Mudminnows are carnivorous, lying in wait for suitable prey, which they seize in a forward rush. Insects, mollusks, and crustaceans are their principal foods.

Collecting: The best way to collect these fishes, which you will usually find sheltering in dense vegetation, is with a minnow seine. You will probably catch many more plants than fish. Another method appropriate for catching mudminnows is a minnow trap set in shallow water with dense vegetation.

Handling: An aquarium containing mudminnows should be quite densely planted. If it is to be a community tank, any other species should be those that are also fond of vegetation for shelter; sunfish or pikes would be good choices.

Mudminnows will accept a wide range of foods, including small earthworms, immature aquatic insects, aquatic worms, brine shrimp, or dried prepared foods.

14. Central Mudminnow
Umbra limi

Field Marks:
· Body is nearly round in cross section near the head, more laterally compressed toward the tail.

Adult Size: 2–4 in. (5.08–10.16 cm) average

Color: Ground color on upper parts of the head and body is medium olive-green. The lower sides, belly, and underparts of the head are pale yellow, ochre, or cream. The back and upper sides are marked with 10–15 indistinct, dusky or dark brown vertical bands. A distinct black vertical bar is at the rear of caudal peduncle, nearly at the base of caudal fin. Fins are all pale olive and are unmarked, except for the caudal fin, which has a few randomly placed dark brown spots.

Habitats: See family Habitats.

Natural History: The Central Mudminnow begins spawning when the water has reached about 55°F (13°C). At this time males and females pair off and move into shallow, weedy, usually flooded areas of the ponds and streams in which they live. The spawning ritual is not elaborate; no nest is built and the eggs are simply released and fertilized and allowed to fall onto the submerged vegetation.

Local Names: Mississippi Mudminnow, Western Mudminnow, Dogfish, Mudfish

15. Eastern Mudminnow
Umbra pygmaea

Field Marks:
· Very similar to the Central Mudminnow, except that the indistinct vertical bars of the Central Mudminnow are replaced by indistinct horizontal stripes running along the sides.

Adult Size: 2–3 in. (5.08–7.62 cm) average

Color: Very similar to the Central Mudminnow.

Habitats: See family Habitats.

Natural History: No studies are currently available; the natural history of this species is presumed to be very similar to that of the family.

■ PIKES

Order Salmoniformes
Family Esocidae

Field Marks:
· Body is distinctly elongated and slender, covered with small but obvious scales and a thick mucous coating. The absence or presence of scales on gill covers and cheeks serves to distinguish one species of the family from another.
· Lateral line is complete but may not be obvious.
· Head is large with sharply pointed snout. Lower jaw protrudes beyond tip of upper jaw. The eyes are proportionately small.
· Mouth is very large, although upper jaw ends near midpoint of eyes. Teeth are large, sharp, and plentiful. Seven soft-rayed fins: paired pectorals set very close to rear edge of gill covers and low on the body; paired ventrals at or slightly behind the midpoint of the fish's total length; a single rounded or squared anal fin

originating far back on the midline of the belly; a moderately to deeply forked caudal fin; and a single rounded dorsal fin directly above the anal fin.

Habitats: All pikes live in shallow weedy waters, most often in large clear lakes, ponds, or slow-moving rivers. Larger pikes may move into deeper, cooler water in the hottest part of summer. Pikes are often found in the same habitats as are members of the sunfish family.

Natural History: Members of the pike family spawn in early spring or very late winter. In regions where lakes, ponds, and streams are frozen during the winter months, spawning usually begins when the ice has just melted. Some species engage in late summer or fall spawning as well. Pikes do not build nests, but simply scatter the fertilized eggs over dense, submerged vegetation. Neither the eggs nor the young receive any parental care.

Both young and adult pikes are carnivores and feed on smaller fishes. They will, however, strike at almost anything of suitable size that moves near them in the water.

Collecting: Since pikes can often be seen lurking near the surface, minnow seines and long-handled dip nets in weedy shallows are good devices for collecting aquarium-sized specimens. Rod-and-reel fishing with small live baits on barbless hooks is another option. Netting an area will usually produce more small pikes than will angling.

Handling: Pikes suffer, sometimes fatally, from rapid changes in water temperature. Both the transporting and aquarium waters should be as close as possible to that of the temperature of their natural habitat.

Aquariums for the pikes should be densely planted with open areas left between plants. The addition of a sunken branch or small tree stump will make the aquarium look even more like the pikes' natural habitat.

Pikes prefer live foods, especially small fishes. Collecting very small fishes is more economical than purchasing them from an aquarium dealer. If live fish are not readily available, brine shrimp or aquatic worms are good alternatives.

The pikes are not ideally suited for prolonged life in an aquarium and should be kept for only a few weeks at a time.

Note: The Redfin Pickerel and the Grass Pickerel are usually considered to be subspecies of the same species. They are given individual attention here because they show some outward differences that may aid the collector in identification.

16. Redfin Pickerel

Esox americanus americanus

Field Marks:

· Body is somewhat more robust than in other members of the family; nearly cylindrical, only very slightly compressed. Body scales are small and numerous.

· Lateral line is distinct.

· Snout is shorter and less pointed than Grass Pickerel's.

· Cheeks and gill covers are scaled.

· Dark band under the eye slants backward.

· Caudal fin is moderately forked.

Adult Size: 6–12 in. (15.24–30.48 cm) average

Color: The back ranges from dark brown or olive to almost black. The young have a pale yellow, unbroken lateral band running along their sides. When fish are 4–6 in. (10.16–15.24 cm) long, this band is gradually broken up by 20–35 olive to dark brown vertical bars. The belly is pale amber to pure white. Dorsal fin is the same dark color as the back, with tinges of amber. Other fins are intense amber or reddish color.

Habitats: The Redfin Pickerel is most common in slow-moving streams where it lies hidden in or near dense vegetation. Most often found in acidic, tea-colored water.

Natural History: The Redfin Pickerel spawns in early spring, from March through May, depending on locality. Recent studies suggest there is also limited spawning in late summer and fall. When this species occurs with the Chain Pickerel, the two may interbreed, producing offspring resembling both parents but growing larger than pure Redfin Pickerel. After hatching, Redfin Pickerel grow very rapidly for the first 2–3 years of their average 5–7-year life span.

Local Names: Banded Pickerel, Mud Pickerel, Grass Pickerel, Bulldog Pickerel, Red-finned Pike

17. Grass Pickerel
Esox americanus vermiculatus

Field Marks:
· The smallest member of pike family, which can be distinguished from the Redfin Pickerel by several traits:
· The snout is much more pointed and longer—more typical of the other members of this family than the Redfin Pickerel.
· The dark band below the eye is nearly vertical.
· Both the cheeks and gill covers of the Grass Pickerel are scaled, as in the Redfin Pickerel.

Adult Size: 6–10 in. (15.24–25.40 cm) average

Color: The Grass Pickerel's color and markings are very similar to the Redfin Pickerel's. The continuous pale yellow lateral band evident in young specimens begins to break up when the fish are about 5 in. (12.70 cm) long. Vertical bars, present in adults of both species, may be more numerous in the Grass Pickerel than in the Redfin. Light-colored areas between bars are generally wider than the bars themselves. Young fish are much darker than adults. An intense ochre stripe runs along the midline of their back, beginning at the top of the snout and ending at the leading edge of dorsal fin. The sides of young specimens are marked with two long dark stripes separated by a wide greenish-yellow band. The belly in all ages varies from buff to white. All the fins are amber to grayish with black leading edges.

Habitats: Nearly always found in shallow water. While the Redfin Pickerel is most common in streams, the Grass Pickerel seems to prefer ponds and small lakes, where it shelters near aquatic vegetation or submerged brush piles. Like the Redfin, the Grass Pickerel prefers acid, rather than alkaline, water.

Natural History: Nearly identical to that of Redfin Pickerel.

Local Names: Little Pickerel, Grass Pike, Mud Pickerel, Mud Pike

18. Northern Pike

Esox lucius

Field Marks:

· Upper parts of gill covers and entire cheeks are scaled; lower portions of gill covers are not scaled.
· Caudal fin is moderately forked with rounded lobes.
· Usually 5 sensory pores under each side of lower jaw; occasionally 3, 4, or 6 may be on one side only.

Adult Size: 8–36 in. (20.32–91.44 cm) average

Color: Both young and adult Northern Pike have similar markings and color. The upper back is dark green or olive over pale yellow or green sides. The belly is white to cream. The sides are sprinkled with pale kidney-shaped spots. The fins may be amber to brick red. All fins (except the pectorals) are marked with wavy dark lines or spots.

Habitats: The Northern Pike is usually found in cool lakes, large ponds, and rivers with weedy shorelines. Aquarium-size pike are usually found in shallow water near shore.

Natural History: A mature female pike may produce from 7000 to 50,000 eggs. The fertilized eggs hatch in 7 to 10 days at a water temperature of about 52°F.

Under favorable conditions, the Northern Pike is one of North

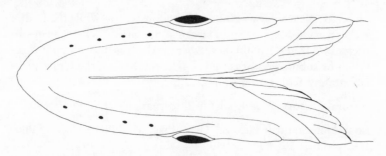

Northern Pike — Underside of Head

America's fastest growing freshwater fish. It has been known to reach a length of 12 in. (30.48 cm) in its first year and 23 in. (58.42 cm) by its third summer. Because of its uncommonly rapid growth rate the Northern Pike will soon outgrow the average home aquarium. A tank with a minimum capacity of 50 gallons is recommended.

The diet of the Northern Pike is primarily made up of smaller fishes; large individuals may eat fish more than 1 ft. (30.48 cm) in length. Like other members of its family, it will consume almost anything that comes within its range.

Local Names: Great Northern Pike, Jack, Jack-fish, Pickerel, Snake

19. Chain Pickerel

Esox niger

Field Marks:

- Cheeks and gill covers are completely scaled.
- Caudal fin is deeply forked.
- Four sensory pores (occasionally 5 on one side only) on the undersurface of lower jaws.
- Dark band below the eye slants slightly backward.

Adult Size: 6–24 in. (15.24–60.96 cm) average

Color: Both juveniles and adults are dark olive, brown, or very dark gray on the back. The young have a distinct yellow stripe along midline of the back. The sides of young fish are marked by indistinct dark vertical bars that develop with age into the typical adult chain pattern. Bellies of both young and adults are creamy-white to dusky. The ground color of the sides in young and adults runs from pale yellow-ochre to greenish-gold. In adults the dorsal, anal, and caudal fins are dark ochre or brown, while the pectorals and ventrals are lighter ochre to yellow. All fins in juveniles are light ochre to greenish-yellow.

Habitats: See family Habitats.

Natural History: In late winter or early spring, large groups of Chain Pickerel move into shallow, often flooded, areas in which they live. A great deal of tail splashing and darting about accompanies the spawning act. No nest is prepared. Fertilized eggs are simply broadcast over the vegetation. The parents desert the area immediately after spawning. The early survival of any year-class of Chain Pickerel depends heavily on the ecological stability of the spawning areas. Unusually hot or dry spring weather may leave eggs and fry stranded, either out of water or in isolated pools with no access to the larger body of water. Chain Pickerel also have a limited spawning period in late summer and fall.

The Chain Pickerel feeds on small fishes, frogs, snakes, and occasional small mammals.

Local Names: Eastern Pickerel, Lake Pickerel, Duck-billed Pike, Green Pike, Picquerelle

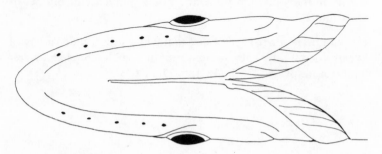

Chain Pickerel — Underside of Head

■ CHARACINS

Order Cypriniformes
Family Characidae

Field Marks:

· Body is strongly compressed laterally, quite deep from dorsal to ventral in most species.
· Body scales are medium-sized to large. No scales on the head.
· Lateral line may be complete or absent.
· Head is small to medium-sized and roughly triangular in profile.
· Mouth is small, terminal.
· Seven or 8 soft-rayed fins (some species lack the adipose fin): paired pectorals set low on the body just behind the gill covers; paired ventrals originating about ⅓ of the total length back from the tip of the snout; a long-based (in most species) anal; a moderately to deeply forked caudal; a small, usually transparent adipose (in most species); and a short-based dorsal of widely varying shape.

Habitats: Characins are found in a wide range of habitats including lakes, ponds, rivers, streams, and springs.

Natural History: Characins spawn at various times, depending on locality; generally in spring and early summer. Their diets include small aquatic insects and plant matter.

Collecting: A minnow seine and a large dip net are the best tools for collecting members of this family. The Characins are popular species in the aquarium trade and can be purchased from pet or aquarium shops.

Handling: These are hardy fishes that will accept a wide variety of live or dried foods. They are best kept in a small school at a water temperature of 65°–70°F (18.48°–21.11°C).

20. Mexican Tetra

Astyanax mexicanus

Field Marks:
- Body is moderately deep.
- Scales are large.
- Lateral line is complete.
- Adipose fin is present.

Adult Size: 3–6 in. (7.62–15.24 cm) average; 6 in. usual maximum

Color: The upper parts of the body and head are light steel-blue, slightly paler below. The lower parts of the head, oval-shaped area on the belly, and lower sides are very pale yellow. The whole fish has a reflective silvery sheen. Dusky spot is just behind the upper part of gill covers and a narrow bluish or dusky lateral band along the sides. Fins are all nearly transparent and unmarked. Anal fin is very lightly tinted with pale sienna or brown.

Habitats: The Mexican Tetra is found in rivers, streams, and springs that remain at a constant temperature.

Natural History: Presumed to spawn in late spring, the Mexican Tetra undertakes short seasonal migrations to avoid the cool wintertime water temperatures encountered in northern parts of its range.

They feed on aquatic or terrestrial insects, aquatic plants, and algae.

■ MINNOWS

Order Cypriniformes
Family Cyprinidae

Field Marks:
- Body is generally elongated, slender (robust in some species), cylindrical or oval in cross section (deep and laterally compressed in a few species).
- In most species the body is covered with well-developed scales.

- Lateral line is usually complete, distinct; in some species it is absent.
- Head small to medium-sized, and generally oval or triangular in profile; tubercles may be present around snout and on the head, especially during spawning seasons. No scales on the head.
- Mouth small to medium-sized, either terminal or inferior. Some species have small barbels at the corners of the mouth. Some have well-developed teeth in the mouth; all species have teeth in the throat.
- Seven soft-rayed fins: paired pectorals set low on the sides, just behind the gill covers; paired ventral fins originating at about midpoint of total length; a single anal fin, usually short-based, squared off or rounded; a broad caudal fin, which is usually moderately to deeply forked; and a single, usually short-based, dorsal fin, either squared or rounded.

Habitats: Members of this large family may be found in nearly all freshwater habitats. The minnows are strictly freshwater fishes.

Natural History: Minnows vary widely in their habits and in the details of their life cycle. Some species construct elaborate nests, while others simply scatter the fertilized eggs over vegetation or the bottom. Minnows generally spawn in spring and most are quite prolific, producing vast numbers of eggs. In many localities minnows are by far the most common species of freshwater fish. They generally provide food for many larger predatory species.

The diets of the minnows are also diverse; foods include aquatic worms and crustaceans, as well as immature and adult insects, aquatic and terrestrial. Some species are bottom-feeders, while others take their food wherever they find it.

Collecting: Minnows are easy to collect with a minnow seine, large or small dip nets, glass or wire minnow traps, or rod and reel.

Handling: Most members of this family are ideal subjects for the beginning fish watcher. They are usually abundant, easy to collect, and generally undemanding in the aquarium. Also, many of the species remain quite small throughout their lives, making them good choices for breeding in the aquarium.

Minnows can be fed a broad range of live or prepared foods. For best results, try to duplicate the feeding habits of the species in the wild as closely as possible.

21. Stoneroller

Campostoma anomalum

Field Marks:
· Body is stout, robust; this min-
  now looks more like a sucker
  than the typical minnow.
· Scales are medium-sized.
· Lateral line is complete.
· Head is large with a squared
snout that slightly overhangs mouth. No scales on the head.
· Mouth is inferior.
· Caudal fin is moderately forked with rounded lobes.
· Breeding males develop prominent nuptial tubercles on their
  head, snout, and dorsal ridge.

Adult Size: 4–7 in. (10.16–17.78 cm) average; 10–11 in. (25.40–27.94 cm) maximum

Color: In females and nonbreeding males, the back and the upper sides are dark olive or brown. Lower sides and belly are pale ochre to silvery-white. The sides are marked with randomly scattered dark spots and sometimes a dark lateral band. Fins are pale tan or yellow. There is a wide dark band on the middle of dorsal fin and a dark triangle at the base of caudal fin. In breeding males, the fins, head, and body are flushed with shades of intense yellow.

Habitats: The Stoneroller is usually found in clear, cool streams with moderate to swift current, over gravel or rubble bottoms in pools, or over riffles. Will tolerate siltation and turbidity. The Stoneroller is usually very abundant within its range.

Natural History: Stonerollers begin preparations for spawning in March and continue spawning into June, depending on location. Mature fish migrate into tributary streams where the males dig shallow pits in the gravel bottom of a riffle. When the nests are completed, the females enter the nesting area and mating begins. With the help of the tubercles on his head and back, the male maintains contact with his mate as they vibrate rapidly together, releasing the eggs and sperm into the nest. Two or more males may simultaneously fertilize the eggs of one female. The male then covers the nest with loose gravel. The fish leave the nest area shortly after spawning.

The Stoneroller spends its life near the bottom, where it feeds on insect larvae, crustaceans, and mollusks.

Local Names: Knottyhead, Hornyhead

22. Goldfish

Carassius auratus

Note: The accidental introduction of domestic Goldfish into widely scattered waters across central and southern North America has resulted in the establishment of thriving wild populations of this exotic species.

Field Marks:
- Body is deep, robust, oval in cross section near the head, more laterally compressed toward the tail.
- Scales are very large, conspicuous.
- Lateral line is absent.
- Small tubercle on each nostril.
- First ray of anal fin and first two rays of dorsal fin are hardened and sawtoothed.
- Caudal fin is moderately to deeply forked, with rounded lobes.

Adult size: 6–12 in. (15.24–30.48 cm) average

Color: Goldfish that have recently been introduced into the wild are usually bright orange or golden. Ensuing generations become darker and begin to take on the bronze or brownish color of the Common Carp. The body and fins are a uniform color, unless the domestic form was variegated, with the caudal and dorsal fins slightly darker than the rest.

Habitats: Goldfish thrive in small bodies of warm water containing plenty of aquatic plants.

Natural History: Wild Goldfish spawn in the spring, usually from March through June, although ripe fish have been taken in the fall, which suggests that spawning may continue throughout the warm months. Eggs are fertilized as they are scattered over dense vegetation; they are adhesive and cling to the plants they touch. Spawning usually takes place on bright, sunny mornings when the fish can be seen splashing and rolling about in shallow water.

Parents desert the area after spawning and the unguarded eggs hatch in an average of 3 days. The young and adults are omnivorous, feeding on aquatic insects, small mollusks and crustaceans, and some plant matter.

Local Names: Golden Carp

23. Rosyside Dace

Clinostomus funduloides

Field Marks:
· Body is moderately deep and slightly compressed laterally.
· Scales are small.
· Lateral line is complete, nearly straight.
· Mouth is large with the upper jaws ending just ahead of the midpoint of the eye.
· Caudal fin is long and deeply forked.

Adult Size: 3–5 in. (7.62–12.70 cm) average

Color: The upper back is dark olive above a narrow stripe of light olive or gold. Below this, there is a broad dark lateral band. The lower sides are pink or rose. The belly is cream or white. The upper back and sides may be lightly sprinkled with small dark spots. Dorsal and caudal fins are olive with light ochre rays. The other fins are transparent and unmarked.

Habitats: Usually found in clear cool streams with gravel bottoms. Most often in shallow water. May also be found in turbid water.

Natural History: As the spring spawning season approaches, male Rosyside Dace claim and defend rather loosely defined territories in shallow riffles of a stream. This territory is abandoned just before spawning, when the males gather in dense groups just downstream from the nests of the Creek Chub. The female Dace remain just downstream from the school of males. Spawning begins when a female moves quickly through the male school, and is followed by one or more males into the nest of a Creek Chub (see pp. 94–95). Eggs are released and fertilized as the pair vibrate in unison. Spawning often occurs while the male Creek Chub is on his nest. Spawning may take place several times in rapid succession, although several minutes normally elapse between releases. Spawning takes place at a water temperature of 65°–68°F (18°–20°C).

Young Rosyside Dace feed on plankton. Adults feed on immature and adult aquatic insects.

Local Names: Red-sided Shiner, Redside Dace

24. Lake Chub

Couesius plumbeus

Field Marks:
· Body is nearly round in cross section.
· Scales are medium-sized.
· Lateral line is complete.
· A small barbel is at each corner of the mouth.
· Caudal fin is long and deeply forked, with rounded lobes.

Adult Size: 3–5 in. (7.62–12.70 cm) average

Color: The back and upper sides range from dark olive or brown to lighter shades of the same colors. Below this is a dark brown or black lateral band. The lower sides are pale cream to silvery-white. The belly is white or silvery. Fins are nearly transparent and unmarked, except for a yellowish tint on the pectorals.

Habitats: The Lake Chub is found in nearly all water types in the northern half of North America. Commonly found in slow-moving streams, ponds, and lakes.

Natural History: Lake Chub often migrate short distances from large bodies of water to reach their preferred spawning beds in clear streams with moderate currents. Spawning may begin as early as April and continue well into August. The Lake Chub spawns freely over the gravel bottom, making no nest and claiming no territory. After spawning, the adults return to the larger body of water.

Young Lake Chub feed on plankton while adults feed on immature and adult aquatic insects and some algae.

Local Names: Northern Chub, Chub Minnow, Plumbeus Minnow, Bottlefish

25. Common Carp

Cyprinus carpio

Field Marks:
· Body is deep, robust, slightly compressed laterally.
· In most forms the body is covered with large scales; some have only a few scales scattered over the body.
· Lateral line is complete, obscure.
· Two small barbels are on each side of upper jaw.
· First ray of the anal fin and the first ray of the dorsal fin are hardened and have a serrated trailing edge.
· Caudal fin is deeply forked with rounded lobes.

Adult Size: 8–36 in. (20.32–91.44 cm) average

Color: The back is very dark brown, dark olive, or black. The sides are golden yellow or bronze with a metallic sheen. The belly varies from cream to pale yellow or amber. All the fins are amber or pale bronze.

Habitats: Common Carp have been introduced to and thrive in a very wide range of North American habitats. Found in lakes, ponds, rivers, and streams, Carp most frequently occur in slow-moving portions of rivers and large streams or the shallow weedy areas of lakes and ponds.

Natural History: The Common Carp, originally found only in Asia and eastern and central Europe, was introduced into North American waters in the late 1800s as a potential food fish.

This is a very prolific species. The average mature female will produce from 35,000 to 2,000,000 eggs. As the spring and early summer spawning season approaches, carp gather in large groups in warm shallow water near shore. Ripe individuals then form smaller groups of 5–15 members. There are usually more males than females in a group. Each female will be accompanied by two or three males during spawning, which includes a great deal of boisterous tail splashing and rolling about. Fertilized eggs are broadcast over submerged vegetation to which they adhere and remain during the 3–6 day incubation period. Spawning may continue over a period of 2–3 weeks at a water temperature of 62°–63°F (17°C). If the water should cool, spawning will be interrupted until the water temperature rises again. After hatching, Common Carp pass through a brief larval stage. Under favorable conditions carp grow quite rapidly and may live for 15–18 years.

Common Carp will eat almost anything they encounter as they browse along the bottom. They use their sucker-like mouths to pick up small portions of bottom sediment, filter out the mud and silt, and consume the edible matter. Their diet usually consists of large numbers of immature aquatic insects, worms, small snails, clams, and shrimp. Carp also eat aquatic plants.

Local Names: German Carp, Mirror Carp, Leather Carp

26. Cutlips Minnow

Exoglossum maxillingua

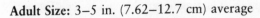

Field Marks:
· Body is stout, tapering little from head to tail.
· Scales are small.
· Lateral line is complete.
· Snout overhangs mouth; the lower jaw is divided into three distinct sections.
· Caudal fin is moderately forked, with rounded lobes.

Adult Size: 3–5 in. (7.62–12.7 cm) average

Color: The back and upper sides are dark brown or olive. The lower sides are silvery-white or cream. The belly is white. All the fins are cream or light gray and are unmarked.

Habitats: Prefers warm clear streams with very little vegetation; usually seeks shelter under stones in quiet water.

Natural History: The Cutlips Minnow is a spring spawner, beginning in May or June and often continuing well into July. The male

Cutlips Minnow — Underside of Head

is responsible for constructing the nest and will aggressively drive away any fish that enter his territory. The nest is a low circular mound of stones with a flat top, about 12–18 in. (30.48–45.72 cm) across and 3–6 in. (7.62–15.24 cm) high. The fish carefully selects each stone and carries it to the nest site in his mouth. The nest mound is usually located near some protective structure, such as a large rock or sunken log, but is always in current sufficient to aerate the eggs. When the nest is finished a female enters the territory and usually is accepted by the male. The pair then move to a position just upstream of the nest and press close together, the female lying just slightly downstream from her mate, while the eggs are released and fertilized. The spawning is accomplished in a series of short bursts, each lasting about 3 seconds. The fish may remain paired through several egg releases or the female may move off to mate with another male. Between spawnings the male continues to work on the nest, adjusting or adding stones. At the completion of spawning most nests contain the eggs of several females, all fertilized by the same male, who remains at the nest site for several days. The fertilized eggs that have slid into the crevices in the mound require about 7 days' incubation period. The fry remain in the shallows for a short period, feeding on plankton and other minute foods near the bottom. Adult Cutlips Minnows become increasingly secretive, spending most of their time under rocks or logs. The principal foods of the adults are tiny mollusks, which they glean from streambed rocks, and a wide variety of immature aquatic insects.

Local Names: Little Sucker, Cutlips

27. Brassy Minnow

Hybognathus hankinsoni

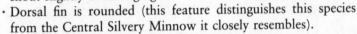

Field Marks:

· Body is slender, oval or cylin-
drical in cross section.
· Scales are medium-sized.
· Lateral line is complete.
· Mouth is subterminal with the
snout slightly overhanging it.
· Dorsal fin is rounded (this feature distinguishes this species
from the Central Silvery Minnow it closely resembles).
· Caudal fin is deeply forked, with rounded lobes.

Adult Size: 2–3 in. (5.08–7.62 cm) average

Color: The back and upper sides are olive or brown. Sides are
brassy yellow with metallic reflections. Belly is dark cream or
white. A dark lateral stripe runs the length of the body, from the
upper rear edge of gill covers to base of the caudal fin. Fins are
nearly clear and unmarked.

Habitats: The Brassy Minnow is usually found in small streams
with clear water and gravel or sand bottoms.

Natural History: The Brassy Minnow is known to spawn in the
spring over a loose gravel bottom in riffle areas of small streams.
Both young and adult Brassy Minnows browse for algae and other
small organisms in the layer of slime and algae covering the bottom of the streams.

Both young and adult Brassy Minnows are found in dense
schools lying close to the bottom.

28. Central Silvery Minnow

Hybognathus nuchalis

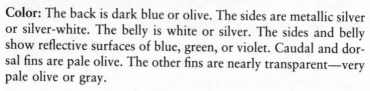

Field Marks:
- Body is elongated and moderately compressed laterally.
- Scales are large.
- Lateral line is complete.
- Mouth is subterminal, slightly overhung by the snout.
- Caudal fin is deeply forked with triangular lobes.

Adult Size: 2–3 in. (5.08–7.62 cm) average

Color: The back is dark blue or olive. The sides are metallic silver or silver-white. The belly is white or silver. The sides and belly show reflective surfaces of blue, green, or violet. Caudal and dorsal fins are pale olive. The other fins are nearly transparent—very pale olive or gray.

Habitats: Found in large schools in shallow weedy areas of large rivers.

Natural History: Spawning begins when the water has warmed to a temperature of about 60°F (16°C). The males are first to move onto the spawning areas while females remain in slightly deeper water. As mating begins, the females move into the shallows where each of them is met by one or more males. The eggs and sperm are released when one or two males press against a female and the pair or group vibrate rapidly over young aquatic vegetation. Other males swim excitedly about the mating fish. Parents desert the eggs at the completion of spawning, with only plants sheltering the incubating eggs. The fry form and maintain small schools for a few weeks after they have hatched. Spawning usually continues throughout the summer months and often into early fall.

The Central Silvery Minnow feeds in the same manner as the Brassy Minnow, consuming algae and very small aquatic invertebrates.

29. Hornyhead Chub
Nocomis biguttatus

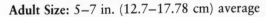

Field Marks:
- Body is robust, resembling the suckers more than the typical minnow.
- Scales are large.
- Lateral line is complete, almost straight.
- Head is large.
- Mouth is slightly subterminal.
- Caudal fin is moderately forked with rounded lobes.

Adult Size: 5–7 in. (12.7–17.78 cm) average

Color: The back and upper sides are olive-brown. A narrow lateral stripe runs from tip of snout to base of tail, where it terminates in a dark spot. Lower sides and belly are creamy or yellowish-white. Young specimens have a bright orange tail that fades as the fish grow older. Breeding males have a bright red spot behind their eyes. Fins are nearly transparent, very pale olive-brown with dorsal and caudal fins slightly darker.

Habitats: Found in clear streams with gravel or rubble bottoms. Adults are usually found near riffles while the young are found in quiet water.

Natural History: The Hornyhead Chub spawns from early spring to early summer, with peak activity in most of its range coming in late May and early June. Most spawning activity takes place after a period of high water.

The male constructs the nest, carrying rocks in his mouth to the chosen site. The finished nest is a small mound of stones, 1–3 feet (.30–.91 m) in diameter and several inches high. Spawning may take place while the nest is being built, in which case, the eggs are deposited into a small depression opened by the male in the larger nest. The nests of Hornyhead and other chubs are often used by other species of minnows. The male Hornyhead Chub tolerates these invasions, appearing oblivious to the intruders as he goes about building the nest, but will drive off other males of his own species. When the nest is completed, a female will enter the territory and the pair will spawn over the gravel mound. As many as 8 or 10 females may mate with one male. The male will continue to add stones to the nest mound after each successive spawning, increasing its size and egg-holding capacity.

The Hornyhead Chub consumes a wide variety of foods, including algae, immature aquatic insects, aquatic mollusks and crustaceans (especially snails), and worms.

30. Golden Shiner

Notemigonus crysoleucas

Field Marks:
· Body is deep and strongly laterally compressed.
· Scales are large.
· Lateral line is complete and curves downward.
· Mouth is very small and terminal.
· Anal fin has a long base and originates far back on the body.
· Caudal fin is long and moderately forked, with triangular lobes.

Adult Size: 3–5 in. (7.62–12.7 cm) average

Color: The back is dark bronze or brown. Sides are golden or brassy. Belly is ochre. Each scale is surrounded by a dark brown or black border, giving the fish a cross-hatched appearance. Body and fins are usually lightly sprinkled with small black spots. The fins are ochre, orange, or light brown.

Habitats: Usually found in warm, shallow, weedy lakes, ponds, or slow-moving streams. Generally travels in loose schools near the water's surface.

Natural History: The Golden Shiner is seldom found at depths greater than 2–3 ft. (.61–.91 m). Its spawning activity usually begins in May and continues through August or September. Spawning takes place as a female and several attending males swim over dense submerged vegetation in shallow water. Eggs are released, fertilized, and allowed to settle into vegetation while the mating fish continue swimming. Fertilized eggs are adhesive. Golden Shiners have been observed using the active nests of Largemouth Bass as repositories for their own eggs. This behavior is remarkable, since Largemouth Bass are known to feed heavily on Golden Shiners when they are available.

The eggs hatch about 4 days after spawning. The fry form dense schools that remain near the surface in shallow water near shore. The young feed mainly on plankton suspended in the water. Adults feed on immature and adult aquatic insects.

Local Names: Butterfish, Chub, Bream, Roach, American Roach, American Bream, Pond Shiner

31. Emerald Shiner

Notropis atherinoides

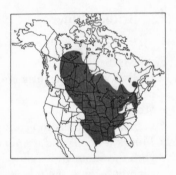

Field Marks:
· Body is long and slender, with a delicate appearance.
· Scales are medium-sized.
· Lateral line is complete and curves downward.
· Anal fin originates under the rear edge of the dorsal fin.
· Caudal fin is long and deeply forked, with triangular lobes.

Adult Size: 2–3 in. (5.08–7.62 cm) average

Color: The back is dusky emerald green. The upper sides are pale metallic green. There is a narrow gray or black lateral stripe. Lower sides are bright metallic silver with white highlights. The belly is silvery-white. All fins are nearly transparent, colorless and unmarked.

Habitats: Dense schools usually found in open water of large lakes and ponds. This shiner characteristically follows the daily rise and fall of plankton, rising toward the surface at dusk, and descending at daybreak.

Natural History: Emerald Shiners may spawn anytime from late May through early July. Spawning takes place after dark, in shallow water over hard bottoms of sand or mud. No nest is prepared, and the nonadhesive eggs simply rest on the bottom during their 2–3 day incubation period. The fry remain near the bottom for several days after hatching and then form dense schools near the surface.

Young Emerald Shiners feed mainly on algae, while adults consume tiny aquatic crustaceans and small aquatic insects.

Special Handling: Emerald Shiners are especially sensitive to abrupt changes in water temperature, so these should be carefully avoided.

Local Names: Lake Shiner, Common Emerald Shiner, Shiner

32. Bridle Shiner

Notropis bifrenatus

Field Marks:
- Body is slender, slightly compressed laterally.
- Scales are large.
- Lateral line is incomplete.
- Mouth is small and terminal with the snout protruding very slightly beyond the upper jaw.
- Anal fin is set far forward on the body, originating below the rear edge of the dorsal fin.
- Caudal fin is long, deeply forked with triangular lobes.

Adult Size: 2–3 in. (5.08–7.62 cm) average

Color: The upper back is black. The sides are light ochre or straw-colored with a prominent black lateral band extending from the snout to the base of the caudal fin. Belly is very pale ochre. Fins are transparent and unmarked.

Habitats: Inhabits a wide range of waters ranging from small, warm-water streams and ponds to large lakes and quiet rivers. Usually found in shallow backwater areas with dense vegetation.

Natural History: The Bridle Shiner's breeding season may last from May through July, beginning when the water temperature has reached 60°F (15°C). Spawning takes place when small groups of mature fish begin swimming over dense vegetation in shallow water near shore. Groups usually include a few females and many more males. Eggs are fertilized as the group keeps swimming along. The fertilized eggs fall into vegetation where they lie for the 2-day incubation period.

The fry remain in the vegetation after hatching. Bridle Shiners are weak swimmers and spend their lives hiding among weeds to avoid predators.

Bridle Shiners feed on plankton and the immature stages of very small aquatic insects, especially midges. They also consume small amounts of algae.

33. Ironcolor Shiner

Notropis chalybaeus

Field Marks:
· Body is elongated, slender.
· Scales are medium-sized.
· Lateral line is complete, nearly straight.
· Snout slightly overhangs the upper jaw.
· Anal fin is set forward on the body, originating below the rear edge of the dorsal fin.
· Caudal fin is long, deeply forked with triangular lobes.

Adult Size: 2–3 in. (5.08–7.62 cm) average

Color: The back is medium to dark brown. The sides are pale tan with a black lateral band running from the tip of the snout to the base of the caudal fin. The belly is pale tan and may have irregularly shaped orange patches between the pectoral and anal fins. The dorsal and caudal fins are pale ochre. The remaining fins are very pale tan (nearly transparent) with tints of yellow or orange.

Habitats: Most often found over sand bottom in bogs and slow-moving streams.

Natural History: Ironcolor Shiners spawn anytime from April to September. Spawning begins when a male vigorously pursues a chosen female, usually throughout the daylight hours. When the female stops swimming she allows her mate to approach and fertilize the eggs as she releases them. No nest is prepared and the fertilized eggs simply settle onto the sandy bottom where they incubate for 2–3 days. The fry absorb their attached yolk sacs in about 5 days, after which they feed on plankton. As adults, Ironcolor Shiners feed on immature and adult stages of both aquatic and terrestrial insects.

34. Warpaint Shiner

Notropis coccogenis

Field Marks:
· Body is elongated, slightly deeper than typical shiner.
· Scales are large.
· Lateral line is complete.
· Caudal fin is deeply forked with triangular lobes.

Adult Size: 3–5 in. (7.62–12.7 cm) average

Color: The back is metallic emerald green. The sides and belly are silvery-white. Above the upper jaw is a narrow orange or red line. Vertical, wedge-shaped, orange or red bar on gill covers. A small, round, red or orange spot is present at the origin of the pectoral fins and a black crescent-shaped mark just behind the gill covers. Dorsal fin is deep yellow with a black band on its top edge. The caudal fin is light tan with a black band on its trailing edge. Other fins are clear and unmarked.

Habitats: Usually found in rivers and large streams with a moderate current, generally over sand or gravel bottom.

Natural History: Just before spawning begins, male Warpaint Shiners claim and defend small territories in the shallow riffles of tributary streams. When mating begins, these territories are abandoned and the mating pairs move to the vacated nests of River

Chub. The female takes up a position slightly downstream from the male. The pair settles into the bottom of the nest, vibrating rapidly as the eggs are released and fertilized. Spawning usually takes place at a water temperature of 70°–80°F (21°–26°C).

Young Warpaint Shiners feed on plankton, while adults feed on immature and adult stages of both aquatic and terrestrial insects.

35. Common Shiner

Notropis cornutus

Field Marks:
· Body is rather stout, robust, moderately compressed laterally.
· Scales are large.
· Lateral line is complete and curves downward.
· Caudal fin is moderately forked, with rounded lobes.

Adult Size: 3–6 in. (7.62–15.24 cm) average

Color: The back is slate blue or olive with some coppery highlights. The sides and belly are silvery. The dorsal and caudal fins are very pale slate gray or olive with orange patches. Other fins, very pale white to almost colorless, may show splashes of orange or yellow.

Breeding males are very brightly colored, with the sides, pectoral, ventral, and anal fins all becoming intense orange or yellow.

Habitats: Most often found in medium-sized streams with a moderate flow over gravel bottom; also occurs in lakes, ponds, and rivers.

Natural History: Studies of stream-dwelling populations of Common Shiners indicate that spawning usually begins in late May or early June and may continue well into July, when the water temperature reaches 60°–65°F (16°–18°C). The males claim and defend small territories on the downstream edge of a shallow riffle. The females form loose schools in deeper water downstream from the males, who may dig shallow small pits in the gravel bottom, but these are neither deep nor elaborate. When their eggs are ripe, the females move upstream and enter the males' territories. During spawning the male wraps his body around the female and forces about 50 eggs at a time out of her, fertilizing them as they are released. One pair of fish may mate several times in succession, or the female will move off and mate with other males. The fertilized eggs are simply scattered over the gravel bottom, deposited in their own nests, or deposited in the nests of other fishes. Eggs become adhesive about 2 minutes after they are fertilized and attach to bottom gravel or are worked into crevices by the current. The eggs hatch in 2–3 days.

Fry feed on plankton. Adults feed on a wide range of organisms, including a large percentage of aquatic insects.

Local Names: Silver Shiner, Redfin Shiner, Creek Shiner, Eastern Shiner

36. Whitetail Shiner

Notropis galacturus

Field Marks:
· Body is robust, slightly compressed laterally.
· Scales are large and appear diamond-shaped.
· Lateral line is complete, curves downward.
· Anal fin is set far forward, originating below the rear of the caudal fin.
· Caudal Fin is deeply forked, with triangular lobes.

Adult Size: 2–3 in. (5.08–7.62 cm) average

Color: The back and upper sides are olive or light brown with a metallic sheen. The lower sides and belly are silvery-white with reflective surfaces of green or brown. A black stripe begins below the front of the dorsal fin and increases in width until it ends at the base of the caudal fin. The lateral line shows up as a series of dark gray or brown spots below this stripe. The fins are all clear and nearly transparent except for a dusky spot at the base of the caudal and white on the outer edges of the base of caudal. A dark line is on the trailing edge of the dorsal.

Habitats: Usually found in loose schools in shallow riffles of clear mountain streams with a moderate current.

Natural History: The spring spawning season begins when male Whitetail Shiners dig shallow nests in the gravel bottom of a shallow riffle. Males aggressively guard the nest and a small surrounding territory. When the nests are completed and the females are fully ripe, a female will enter each of the nests. Spawning usually takes place in late morning and early afternoon at a water temperature of about 75°F (30°C). During spawning both fish turn onto their sides as they simultaneously release the eggs and sperm. This process is repeated several times as the eggs are released in short bursts. The fertilized eggs are adhesive.

The young feed on plankton and then graduate to a diet of aquatic and terrestrial insects.

37. Spottail Shiner

Notropis hudsonius

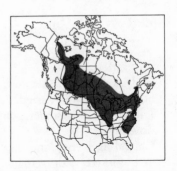

Field Marks:
- Body is elongated, robust, moderately compressed laterally.
- Scales are large.
- Lateral line is complete, nearly straight.
- Snout very slightly overhangs the small mouth.
- Caudal fin is long and deeply forked, with triangular lobes.

Adult Size: 2–3 in. (5.08–7.62 cm) average

Color: The back and upper sides are pale emerald green to light brown above a medium gray or light brown lateral stripe. Below this is a broad pink lateral band bordered on its lower edge by a thin gray lateral stripe. The sides and belly are silvery-white.

There is a small dark spot at the base of the caudal peduncle. Fins are clear and unmarked except for occasional tinges of pale gray near their bases and at their tips.

Habitats: Found in nearly all North American water types from brackish coastal streams to lowland ponds and streams to mountain lakes and streams. Usually found in shallow water over sandy or rocky bottom with little vegetation.

Natural History: The Spottail Shiner usually spawns over gravel shoals in larger lakes and rivers or in the lower portions of tributary streams. Spawning takes place in spring and early summer, beginning in May in southern populations and as late as July in the northern part of its range. Little is known of the spawning habits of this fish. It is thought to spawn far from shore, making it difficult to observe. It is a prolific species, with the average 2-year-old female producing some 1500–2500 eggs.

Young Spottail Shiners feed on plankton. Adults feed on aquatic insects, algae, and small aquatic crustaceans.

Local Names: Spottail, Spottail Minnow

38. Rosyface Shiner

Notropis rubellus

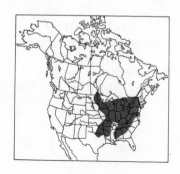

Field Marks:
· Body is elongated, slender.
· Scales are medium-sized.
· Lateral line is complete, curves downward.
· Caudal fin is long and moderately forked, with triangular lobes.

Adult Size: 2–3 in. (5.08–7.62 cm) average

Color: The back is light gray, light olive, or light brown. Ground color of sides and belly is silvery-white. Sides may be marked with a broad pink or lavendar lateral band bordered on the bottom edge of its rear half by a thin grayish-blue stripe. Fins are all nearly transparent and are unmarked.

Habitats: Most often found in the lower reaches of rivers and streams with a moderate current. Occasionally found in lakes and ponds, but nearly always near the mouth of a tributary stream or river.

Natural History: Rosyface Shiners begin spawning when the water temperature approaches 80°F (27°C). The mature fish seek out shallow riffles near the mouths of streams for nesting sites. When the spawning area has been selected, groups of about 10–12 males and females gather together, males far outnumbering females in any school. No territories are claimed and the males show little, if any, aggression toward one another. Spawning usually takes place over the nests of other species of fish, including the Hornyhead Chub, the Stoneroller, the Chestnut Lamprey, and the Longnose Gar. The eggs and sperm are simultaneously released as the fish in the mating school vibrate rapidly. The parents desert the spawning area shortly after the completion of spawning.

Immediately after hatching, the fry wriggle into crevices in the gravel bottom, where they remain for several days. The young feed on plankton, while the adults feed on caddis fly larvae and other small aquatic insects.

39. Spotfin Shiner

Notropis spilopterus

Field Marks:
· Body is elongated, moderately robust.
· Scales are large.
· Lateral line is complete, nearly straight.
· Anal fin is set far forward, originating under rear edge of the dorsal fin.
· Caudal fin is long and deeply forked, with triangular lobes.

Adult Size: 2–3 in. (5.08–7.62 cm) average

Color: The back is a dark steel blue or slate gray. The sides are light slate gray or steel blue with a broad dark gray lateral band that is bordered on its top and bottom edges by thin whitish lateral stripes. The belly is silvery-white. The dorsal and caudal fins are pale gray. The rear portion of the dorsal fin is marked with a distinct black spot. Remaining fins are nearly transparent with flushes of yellow.

Habitats: Spotfin Shiners are found over sand or gravel bottom in large slow-moving rivers and occasionally in lakes. They tolerate considerable turbidity.

Natural History: Typical of its family, the Spotfin Shiner begins spawning in May, but may continue well into August. Prior to spawning, the males claim small territories on parts of logs or tree roots in moderate to swift current. Females remain in the current downstream from the males. As they are ready, individual females will move into the spawning sites and mate with a male. The fertilized eggs are adhesive and attach to the structure over which they were laid.

Young fish feed on plankton and adults feed on aquatic insects.

Local Names: Satin-finned Minnow, Silver-finned Minnow

40. Sand Shiner

Notropis stramineus

Field Marks:
· Body is slender, delicate with a peak at the origin of dorsal fin.
· Scales are medium-sized.
· Lateral line is complete, curves downward.
· Dorsal fin is placed unusually far forward.
· Caudal fin is long and moderately forked, with sharply triangular lobes.

Adult Size: 2–3 in. (5.08–7.62 cm) average

Color: The back and upper sides are light reddish-brown or ochre. A short brown lateral stripe below the dorsal fin extends toward the tail; below this stripe, the sides are very pale ochre. The belly is creamy-white or straw. The scales are surrounded by a dark border, giving the fish a cross-hatched appearance. Fins are all very pale tan and are unmarked except for a small brown spot on the base of the caudal fin.

Habitats: The Sand Shiner prefers streams with little or no vegetation over sand or gravel bottom. Also found in shallow parts of lakes and ponds over sand or gravel bottoms.

Natural History: What little information is available suggests that the Sand Shiner begins spawning in May or June and may continue throughout the summer. The Sand Shiner builds no nest. Eggs are laid and fertilized over gravel bottoms of streams or on shoals in lakes or ponds.

Adults are usually found in loose schools, feeding on aquatic or terrestrial insects at or just below the surface.

Local Names: Shore Minnow, Straw-colored Minnow, Northern Sand Shiner

41. Mimic Shiner

Notropis volucellus

Field Marks:
- Body is slender, slightly compressed laterally with a slight hump that peaks at the origin of the dorsal fin.
- Scales are medium-sized.
- Lateral line is complete and appears as a series of dark X-shaped spots below the lateral band.
- End of the snout is blunt, rounded.
- Caudal fin is long and moderately forked, with triangular lobes.

Adult Size: 2–3 in. (5.08–7.62 cm) average

Color: The back and upper sides are medium brown to dark ochre. A distinct dark brown, gray, or black lateral band runs from the rear of the gill covers to the end of the caudal peduncle.

Below the lateral band, the sides are pale tan or ochre with a metallic sheen. The belly is cream or straw-colored. Fins are light tan and are unmarked. Pectoral, ventral, and anal fins may be flushed with pale yellow.

Habitats: The Mimic Shiner occurs in lakes and ponds, where it remains in shallow water close to shore during the day and moves to deeper water at night.

Natural History: Probably spawns after dark in deep water. Eggs are fertilized as they scatter over vegetation. Mimic Shiners spawn later than most other species of this family, usually reaching their peak in July in most locations.

Adult Mimic Shiners feed on small aquatic crustaceans (such as daphnia), aquatic insects (especially still water midges), and algae.

42. Bleeding Shiner

Notropis zonatus

Field Marks:
· Body is slender, slightly compressed laterally.
· Scales are medium-sized.
· Lateral line is complete, curves downward.
· Anal fin is broad-based, set far forward, originating under the rear edge of the dorsal fin.
· Caudal fin is moderately forked, with rounded lobes.

Adult Size: 3–5 in. (7.62–12.7 cm) average

Color: In females and nonbreeding males the top of the back is very dark, often black, forming a middorsal stripe. The upper sides are golden or ochre. A dark brown or dusky lateral stripe begins as a narrow black band on the head and ends at the rear edge of the caudal peduncle. The lateral line appears as a series of small dark spots, forming a distinct line. The lateral stripe forms a dark patch on the sides ahead of the dorsal fin. The fins are clear, whitish. The trailing edge of the dorsal fin is marked with a whitish band bordered on its inner edge by a dusky band. The caudal fin has a dark patch on its center. The fins and belly of breeding males are flushed with bright red, orange, or yellow.

Habitats: The Bleeding Shiner is often the most abundant fish in pools or quiet portions of riffle areas in clear mountain streams within its range.

Natural History: Schools of brightly colored breeding males may be so abundant within the species' range that they create a red patch on the stream bottom. Spawning usually begins in late April, peaks in late May or June, and is complete by mid-July. The males gather at the head of a riffle and remain close to the bottom. Females take a position downstream from the males but at a higher level in the water. Males often dig shallow pits and engage in jostling and butting to determine dominance and win preferred spots on the spawning site. Nests of Stonerollers and Hornyhead Chub are often used by the Bleeding Shiner. When they are ready to mate, females drop down and move forward, each taking a position close to a dominant male. The male maneuvers the female into a vertical stance and then wraps his body around hers while eggs and sperm are released. After spawning, the female darts straight toward the surface, rejoining the group of females. This process continues until the fish are spent.

The Bleeding Shiner's fertilized eggs sink to the riffle bottom and lodge in loose gravel where they lie for several days before hatching.

Fry feed on plankton and adults feed mainly on aquatic and terrestrial insects.

The Bleeding Shiner often utilizes the same nesting sites as the Striped Shiner, Rosyface Shiner, Redbelly Dace, and Ozark Minnow. This proximity leads to the birth of a number of hybrid individuals.

43. Northern Redbelly Dace

Phoxinus eos

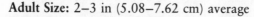

Field Marks:

· Body is elongated, round in cross section with a slight hump that peaks at the origin of the dorsal fin.
· Scales are very small, making the fish appear scaleless.
· Lateral line is incomplete.
· Mouth is small, terminal, ending at the front edge of the eye.
· Caudal fin is long and moderately forked, with rounded lobes.

Adult Size: 2–3 in (5.08–7.62 cm) average

Color: The upper back is dark brown or dark olive to almost black. The sides and belly are pale yellow to brilliant yellow or red. There are 2 dark lateral stripes along the sides, with a dark mottled pattern between them. The fins are yellow, pale tan, or clear. The dorsal and caudal fins may have a few small brown spots or areas of darker color on them. Other fins are unmarked.

Habitats: Most often found in boggy, spring-fed ponds and slower portions of small streams. Often associated with Brook Trout.

Natural History: Northern Redbelly Dace usually begin to spawn in May and often continue sporadically through August. A water temperature of 70°F (21°C) triggers spawning. Individual females dart into a dense mass of algae, followed by one or more males. Eggs are released and fertilized while a female and one or more males thrash about in the algal mass. Eggs are not adhesive and simply settle into the algae, where they remain for the duration of their 8–10 day incubation period.

Fry feed on plankton and algae while adults feed on aquatic insects.

Local Names: Redbelly Dace, Yellowbelly Dace

44. Southern Redbelly Dace

Phoxinus erythrogaster

Field Marks:
· The Southern Redbelly Dace is nearly identical to the Northern Redbelly Dace. Their color and range, except where they overlap, are their major differences.

Adult Size: 2–3 in. (5.08–7.62 cm) average

Color: The back is brown to dark brown or olive. There is a series of small black dots on the upper back just ahead of the dorsal fin. Two narrow dark lateral stripes are separated by a light-colored band, usually pink. Below the stripes, the sides are white, very pale pink, or orange. The belly is pink or orange. The dorsal and caudal fins are light brown. Other fins are pale pink, orange, or white.

Habitats: Occurs in small brooks and streams. Most often found sheltering beneath undercut banks.

Natural History: See Northern RedBelly Dace (p. 86).

Local Names: Redbelly, Yellow Belly Dace

45. Finescale Dace

Phoxinus neogaeus

Field Marks:
· Body is robust, relatively stout, oval in cross section.
· Scales are very small, making the fish appear scaleless.
· Lateral line is incomplete, ending above the origin of the ventral fins.
· Caudal is broad and moderately forked, with rounded lobes.

Adult Size: 2–3 in. (5.08–7.62 cm) average

Color: The back is dark brown to black. The sides are pale yellow, tan, or amber. A single broad dark gray or black lateral band begins at the tip of the snout and ends at the fork of the caudal fin. The belly is white or silvery. In breeding males, the sides below the lateral band are bright yellow or red. The dorsal and caudal fins are light brown. Other fins are pale ochre or amber. The fins are unmarked.

Habitats: Most often found in boggy, peat-stained ponds and small lakes.

Natural History: The Finescale Dace, a spring spawner, usually mates at the same time as the Northern Redbelly Dace with which it is often associated. Similar in spawning behavior, the two species are known to crossbreed, producing fertile hybrid offspring. The spawning period lasts from late April well into May or June. Eggs are laid and fertilized on undersides of sunken logs or other overhanging structures.

Fry feed on plankton and algae. Adults feed on aquatic invertebrates, mollusks, crustaceans, and insects.

Local Names: Fine-scale Minnow, Bronze Minnow

46. Bluntnose Minnow

Pimephales notatus

Field Marks:
· Body is robust, relatively stout, round in cross section.
· Scales are medium-sized, prominent.
· Lateral line is nearly complete.
· Snout is distinctly rounded.
· Caudal fin is moderately forked, with rounded lobes.

Adult Size: 2–3 in. (5.08–7.62 cm) average

Color: The back is dark gray, brown, or olive. The sides are metallic whitish-silver. The belly is white or silvery. A dark gray or black lateral stripe begins at tip of snout and ends at the rear edge of the caudal peduncle, where it forms a dark spot. Dark skin pigment showing around the edges of the scales gives the body a cross-hatched appearance. The fins are light gray or off-white. There is a dark spot on the leading edge of the dorsal fin.

Habitats: The Bluntnose Minnow occurs in a wide range of habitats, including medium-sized rivers, ponds, and lakes. It is tolerant of adverse conditions such as turbidity and pollution.

Natural History: The Bluntnose Minnow begins its spawning season when the water temperature has risen to 70°F (21°C). The male selects a rock or other large structure and clears a spot on its underside by removing small stones and other debris with his tubercled head and snout. He completes the excavation with powerful sweeps of his tail. If a female or other species of fish approaches the nest site before it is completed, the male will aggressively drive them away. When the nest is finished, a female will approach and be accepted by the male. The pair turn upside-down under the overhanging surface and simultaneously release the eggs and sperm. Fertilized eggs adhere to the overhang. Preferred nesting areas may be used by many mating pairs. Nests take up as little as 1 ft. of available space and usually contain eggs in various stages of incubation. Each fish may take several mates during the extended breeding season lasting from May through August. The male remains on or near the nest as long as it contains eggs, keeping them free of silt and other debris by fanning them with his fins. The fry grow very rapidly after hatching.

Bluntnose Minnows consume large quantities of detritus, which has a bacterial coating containing very small invertebrates that nourish the fish.

47. Fathead Minnow

Pimephales promelas

Field Marks:
- Body is elongated, robust.
- Scales are medium-sized.
- Lateral line is nearly complete.
- Head is flattened on top with a spongy pad in the area from the back of the head to origin of the dorsal fin.
- Dorsal fin is distinctly rounded.
- Anal fin is set far forward.
- Caudal fin is long and moderately forked, with rounded lobes.
- Female has a conspicuous ovipositor.

Adult Size: 2 in. (5.08 cm) average

Color: The back is dark olive or dark brown. The sides are light green or yellowish. A dark blue to black lateral stripe runs from the rear edge of the gill covers to the rear edge of the caudal peduncle. This is bordered on its top edge by a thin light lavender stripe. The belly is white or cream. The dorsal and caudal fins are very pale olive or light brown. Other fins are pale ochre or pale yellow. The fins are unmarked.

Habitats: In northern parts of its range, the Fathead Minnow inhabits boggy, often tea-colored lakes, ponds, and slow-moving streams. In the southern ranges it lives in silt or mud-bottomed lakes, ponds, and slow-moving rivers and streams.

Natural History: For many years the Fathead Minnow has been artificially propagated as a forage fish for larger game fishes.

When the water temperature has reached about 65°F (18°C), male Fathead Minnows select nesting sites on the undersides of overhanging objects, such as logs or stump roots and, occasionally, the undersides of lily pads. When the male has finished clearing the nest site, he seeks out a ripe female and prods her into his territory. Using the pad on his head and his well developed nuptial tubercles he pushes his mate under the overhanging structure. As he moves her into position, she turns onto her back and he takes a position beside her. The female's ovipositor helps her effectively direct the eggs upward onto the overhang while the male fertilizes them. After spawning, the male drives his mate from the nest, usually seeks out and mates with several additional females, and jealously guards the nest, which normally contains eggs in various stages of incubation. Eggs usually require about 5 days to incubate fully. Newly hatched fry are translucent white. Fathead Minnows grow very fast under favorable conditions and may reach sexual maturity and spawn at the end of their first summer. Male Fathead Minnows grow faster and attain greater lengths than females. Most individuals die after spawning. The average life span is about 2 years.

Fathead Minnows feed extensively on algae and detritus, but may occasionally take small, immature aquatic insects.

Local Names: Northern Fathead Minnow, Blackhead Minnow

48. Blacknose Dace
Rhinichthys atratulus

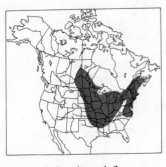

Field Marks:
· Body is elongated, slender.
· Scales are very small.
· Lateral line is complete, nearly straight.
· Snout is pointed and slightly overhangs the mouth.
· Anal fin originates below the rear edge of the dorsal fin.
· Caudal fin is moderately forked, with rounded lobes.

Adult Size: 2–3 in. (5.08–7.62 cm) average

Color: The back is dark brown or dark olive. The sides are very pale greenish or yellowish. A single dark (dusky or black) lateral band runs from tip of snout to midpoint of caudal fin. The sides above and below this lateral band are mottled or lightly spotted with dusky or brown pigment. Fins are light tan or light ochre. All fins are unmarked except for the dark patch on caudal.

Habitats: The Blacknose Dace is most often found in small swift streams with gravel bottom; less frequently encountered in lakes or ponds.

Natural History: When the water temperature has reached 70°F (21°C), male Blacknose Dace move into shallow riffle areas of small streams to claim spawning territories. Once a site has been selected, its tenant aggressively guards against all intruders. These territories are maintained only until spawning begins, then abandoned. As spawning begins, one to four males closely accompany a single female, appearing to harass her. The males clear silt from potential nesting sites although no nest is actually built. Eggs may or may not be deposited in these cleared areas, which apparently are not necessary to successful spawning. During the spawning act a male and female press close together, maintaining contact with the aid of the many nuptial tubercles covering the male's head and snout. The mating pair vibrate together while the eggs and sperm are released. As soon as the eggs are laid, other Blacknose Dace rush in and attempt to eat the eggs. In most cases the male fiercely defends the eggs. The incubation period is not known. Fry feed on microorganisms suspended in water. The bulk of the adults' diet is made up of immature aquatic insects.

Local Names: Brook Minnow, Striped Dace, Eastern Blacknose Dace, Potbelly

49. Longnose Dace
Rhinichthys cataractae

Field Marks:
- Body is elongated, robust, possibly potbellied in some individuals.
- Scales are very small.
- Lateral line is complete, nearly straight.
- Snout is long, pointed, slightly overhanging the mouth.
- A small barbel is hidden in the skin fold at each corner of the mouth.
- Caudal fin is moderately forked, with triangular lobes.

Adult Size: 3–4 in. (7.62–10.16 cm) average

Color: The back is dusky to black. The upper sides are pale ochre or golden. The lower sides are very pale tan or ochre. A dark lateral band runs from the tip of the snout to the rear edge of the caudal peduncle. The upper back and areas adjoining the lateral band are mottled with dark brown or dusky pigment. The belly is cream. The base color of the fins is a very pale tan to almost clear. The dorsal fin has a light brown leading edge. The caudal fin has light brown edges and a dark patch (a continuation of the lateral band) at its base. The pectoral fins have light brown leading edges.

Habitats: The Longnose Dace is usually encountered in schools around shallow riffles in turbulent streams and small rivers. Rarely found in lakes or ponds.

Natural History: The Longnose Dace begins its spawning season when the water temperature has reached about 55°F (13°C), from May to early July. The spawning act is nearly identical to that of the Blacknose Dace (see pp. 91–92). The adhesive eggs simply fall and lodge in the loose gravel. The incubation period is 7–10 days. Fry swim to the surface, where they live for about 4 months feeding on microorganisms. After this, they spend their lives at or near the gravel bottom. The flattened head and streamlined body of the Longnose Dace help maintain its position near the bottom in fast-moving water.

Adults feed on small immature aquatic insects, such as midges and blackflies.

50. Creek Chub
Semotilus atromaculatus

Field Marks:
- Body is slender, slightly compressed laterally.
- Scales are small.
- Lateral line is usually complete, but may be interrupted.
- Snout is pointed with upper jaw protruding slightly beyond the lower jaw.
- A small barbel is concealed at each corner of mouth; may be absent in young specimens.
- Caudal fin is long and moderately forked, with rounded lobes.

Adult Size: 3–5 in. (7.62–12.7 cm) average

Color: The back is dark brown, bronze, or olive. The sides are pale tan or ochre. A broad, dark lateral band begins at tip of the snout and ends at the rear edge of the caudal peduncle. This band fades slightly with age. The belly is cream or white. The dorsal and caudal fins are light ochre or light amber. There is a dark spot at the forward base of the dorsal fin. Other fins are nearly clear and flushed with light yellow or orange. The caudal fin may have light brown patches near base.

Habitats: Usually found in small brooks and streams with cool, clear water.

Natural History: The Creek Chub may well be the best-known North American minnow. Used extensively as a bait fish, its natural history has received much attention from ichthyologists. It begins spawning when the water has reached a temperature of 55°F (12°C). Males select spawning sites just above or below shallow riffles with clean gravel bottoms. The male digs a shallow trench in the bottom by swimming rapidly on his side and carrying stones upstream in his mouth, placing them at the upstream edge of the trench. The female moves onto the completed nest site and is immediately approached by the male, who places his heavily tubercled head and one pectoral fin under the female, then lifts her into an upright position while wrapping his body around hers. During this embrace, the female releases about 50 eggs, which the male fertilizes as they fall into the trench. After each spawning the female floats belly up for a moment, then rights herself and begins swimming again. She may mate several times with the same male or move off and find other mates. The male covers the eggs with rubble from the downstream side of the nest, and digs another nest in preparation for another mating. This process continues until all the ripe females are spent.

In several days the eggs hatch into very fast growing fry, which feed on plankton. As adults, Creek Chub feed on a wide variety of foods, including immature and adult stages of both aquatic and terrestrial insects, other aquatic invertebrates, worms, and some small fishes.

Local Names: Brook Chub, Horned Dace, Common Chub, Tommycod, Mud Chub

51. Fallfish

Semotilus corporalis

Field Marks:
· Body is elongated, robust, oval in cross section.
· Scales are large, conspicuous.
· Lateral line is complete, curves downward.
· In adults a small barbel is concealed in the fold of the upper jaw at each corner of the mouth.
· Caudal fin is long and moderately forked, with triangular lobes.

Adult Size: 4–10 in. (10.16–25.4 cm) average

Color: The back varies from light brown to olive to almost black. The sides are metallic silver or silvery-white, sometimes with a bronze tint. The belly is white with a pearly sheen. The base of each scale is marked with a dark triangle, producing a cross-hatched appearance. Dorsal and caudal fins are light gray, olive, or ochre. Other fins are pale whitish and nearly transparent.

Habitats: The Fallfish is most often encountered in cool, clear lakes, ponds, rivers, and large streams. It generally prefers larger bodies of water than do the smaller minnows.

Natural History: The Fallfish begins spawning when water reaches a temperature of 50°–55°F (10°–13°C). The male selects a nesting site in a shallow riffle and begins building a nest mound by carrying stones in his mouth and depositing them on the nest site. A 7–8-in. (17.78–20.32-cm) fish is capable of erecting a mound 3 ft. (.91 m) across and 2 ft. (.61 m) high.

Spawning commences when a ripe female enters the male's territory after the nest mound is completed. He prods her onto the nest, where the eggs are released and fertilized, and they fall into crevices in the mound. After each spawning the male covers the mound top with a layer of gravel. The Fallfish's nest mound may be used by Common Shiners (p. 77), resulting in hybrid off-spring.

Eggs remain in the mound for a few days before hatching. After hatching, the young pass through a short larval stage in the nest mound. Once they leave the nest mound, the fry feed on plankton. Adults feed mainly on immature and adult aquatic insects, terrestrial insects, and occasional small fishes.

Special Handling: Captive Fallfish will readily eat other fishes, and should not be kept in tanks containing specimens much smaller than themselves.

Local Names: Windfish, Chivin, Chub, American Chub

52. Pearl Dace

Semotilus margarita

Field Marks:

- Body is moderately robust, nearly round in cross section.
- Scales are small.
- Lateral line is complete.
- Dorsal fin originates behind the origin of the ventral fins.
- Caudal fin is moderately forked, with triangular lobes.

Adult Size: 3–5 in. (7.62–12.7 cm) average

Color: The back is dark brownish-gray to nearly black. The sides are the color of tarnished silver. The belly is white or silvery. Individuals may either be heavily spotted or mottled with dark brownish-gray or black or have only a few scattered spots. A diffuse dark lateral stripe runs from just behind the gill covers to the rear edge of the caudal peduncle. The gill covers are often marked with dark patches. In young specimens the lateral stripe may terminate in a distinct dark spot at the base of the caudal fin. All fins are light gray and are unmarked.

Habitats: The Pearl Dace prefers cool, boggy, spring-fed ponds, slow-moving streams, and small lakes.

Natural History: Male Pearl Dace become brightly colored and claim spawning territories when water reaches 65°F (18°C). The territories average about 8 in. (20.32 cm) in diameter and are well separated from each other. Males guard these territories when they are over them, but will often leave them. Mating begins when a male prods a female of his choosing onto his territory. He places his enlarged pectoral fin, covered with nuptial tubercles, under the rear portion of his mate's body and curls his caudal fin over hers. While in this embrace, both fish vibrate rapidly as eggs and sperm are released. A female often moves from one mate to another, releasing small numbers of eggs with each until she is spent.

Young Pearl Dace grow very rapidly, with females generally outpacing the males. Fry feed on plankton. Adults feed mainly on aquatic crustaceans and small immature aquatic insects. They also consume small amounts of algae.

Local Names: Northern Dace, Northern Minnow, Nachtrieb Dace

■ SUCKERS

Order Cypriniformes
Family Catostomidae

Field Marks:

· Body is usually elongated, round or oval in cross section (strongly compressed laterally in some species).
· Scales small to large.
· Lateral line is usually complete, absent in some species.
· Head is medium-sized, oval or triangular in profile, with a blunt rounded or squared snout overhanging the mouth.
· No scales on the head.
· Mouth is inferior with large fleshy lips that are prostrusible, allowing these fish to suck foods up off the bottom.
· Seven fins: paired pectorals set low on the body just behind the gill covers; paired ventrals, usually originating slightly behind the midpoint of the fish's total length; a single, usually short-based anal fin; a long, broad forked caudal fin; and a single, usually squarish, dorsal fin. All fins have soft rays and no spines.

Habitats: Suckers are found in nearly all types of North American freshwater habitats. They inhabit clear lakes, ponds, rivers, and small streams and may be found in still, slow-moving, or swift waters. They occur in either cold- or warm-water habitats.

Natural History: Suckers spawn in spring or early summer, often migrating considerable distances upstream into small tributaries to reach their preferred spawning sites. Suckers do not build nests. They simply scatter their eggs at random, usually over a clean gravel bottom.

All the suckers are bottom dwellers, sucking in small organisms as they browse among rocks and other objects on the bottom. Immature aquatic insects and crustaceans make up the bulk of their diet.

Collecting: Small suckers are most easily caught in minnow traps set in small streams where they spend their early years or in seines fished along the bottom of streams or ponds. Dry dog or cat food makes a good bait. You can catch them with baited hooks but be careful to set the barbless hook as soon as a strike is felt to prevent the fish from swallowing the bait.

Handling: Suckers are undemanding in an aquarium, requiring only appropriate food and clean water. They can be fed small earthworms, immature aquatic insects, or aquatic worms such as tubifex. They also may be conditioned to accept dry prepared foods.

53. Longnose Sucker

Catostomus catostomus

Field Marks:
· Body is slender, nearly round in cross section.
· Scales are small, crowded toward the head.
· Lateral line is complete but not well defined.
· Anal fin is very long.
· In breeding males, the anal fin, the bottom edge of the caudal fin, and the head are covered with nuptial tubercles.
· Caudal fin is slightly forked with triangular lobes.

Adult Size: 6–12 in. (15.24–30.48 cm) average

Color: Adults are usually dark brown or dark gray on the upper back and sides, sometimes with a reddish or steel-blue tint. The lower sides are white or cream, in distinct contrast to the back. The dorsal, caudal, and anal fins are dusky, occasionally with a pale red border. The pectoral and ventral fins are brown or amber, may be flushed with pale pink.

Young specimens (as illustrated) are pale tan or amber with the upper sides and back densely mottled with dark brown. On the lower sides, the mottling becomes more sparse and irregular. The belly and under parts of the head are pale ochre to cream or white. The fins are pale tan or amber; all but the anal fin are marked with irregular, small, dark brown spots. The anal fin is unmarked.

Habitats: The Longnose Sucker is nearly always found in clear, cold waters of deep lakes and ponds. It is occasionally encountered in small tributary streams of large lakes and has been reported from brackish water in regions near the Arctic Circle.

Natural History: The Longnose Sucker begins its spawning season when water temperature has reached 40°F (4.5°C). Where accessible, small tributary streams are used for spawning; otherwise, spawning occurs in the shallows of lakes and ponds. Several males fertilize the eggs of one female as the entire group splashes and rolls about near the surface. No nest is built and the eggs, which become adhesive when fertilized, are broadcast over the bottom. Eggs usually lodge in crevices, where they remain for an average of 10 days before hatching.

Young fish feed on tiny aquatic invertebrates. Adults feed on immature aquatic insects and other invertebrates.

Local Names: Black Sucker, Finescale Sucker, Red Sucker, Sturgeon Sucker

54. White Sucker

Catostomus commersoni

Field Marks:
· Body is slender and oval in cross section.
· Scales are medium-sized, crowded toward the head.
· Lateral line is complete but obscure.
· Pectoral fins are large.
· Anal fin is long.
· Caudal fin is broad, moderately forked.
· In breeding males, the rays on the anal fin and the lower parts of the caudal fin are covered with nuptial tubercles.

Adult Size: 5–18 in. (12.7–45.72 cm) average

Color: Despite their common name, White Suckers vary widely in color according to their habitat. Usually the head and upper back are dark brown, dark olive, copper, or black. The sides are sandy brown, straw-colored, or copper. The lower sides and the belly may be pale ochre, pale yellow, or creamy-white. Young specimens are vaguely mottled with dark brown or gray above the lateral line. Breeding males have intense pink or red bands along their sides and black backs. All fins are light ochre or light brown.

Habitats: Usually found in warm shallow lakes and ponds or in the shallow bays of large, cold-water lakes; also common in slower portions of large rivers and major tributaries of lakes and ponds.

Natural History: Lake and pond populations move into shallow waters near shore or into tributary streams at spawning time, usually returning to the stream of their birth. Stream dwelling populations move upstream and into tributaries. Nearly all spawning activity takes place after dark. Suitable spawning streams may host runs of many thousands of fish. During the day fish are found in large dense schools in deep pools, often just below a riffle. Spawning takes place in shallow gravel-bottomed riffles with several males attending each female. Any number of attendant males may fertilize the eggs. The fish splash and roll about during spawning, an activity that lasts 3–4 seconds and may be repeated as often as 40 times per hour. White suckers are very prolific. Mature females produce as many as 20,000–50,000 eggs. The incubation period lasts an average of 2 weeks at a water temperature of 60°F (15°C). Only about 3%–5% of the eggs survive to the fry stage.

Local Names: Common Sucker, Coarse-scaled Sucker, Fine-scaled Sucker, Mullet, Gray Sucker, Eastern Sucker, Black Sucker, Black Mullet

55. Northern Hog Sucker

Hypentelium nigricans

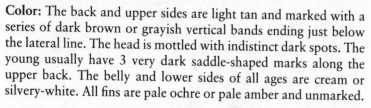

Field Marks:
- Body is elongated, robust.
- Scales are medium-sized.
- Lateral line is complete but not well defined.
- Pectoral fins are noticeably large.
- Caudal fin is broad, moderately forked with rounded lobes.

Adult Size: 5–10 in. (12.7–25.4 cm) average

Color: The back and upper sides are light tan and marked with a series of dark brown or grayish vertical bands ending just below the lateral line. The head is mottled with indistinct dark spots. The young usually have 3 very dark saddle-shaped marks along the upper back. The belly and lower sides of all ages are cream or silvery-white. All fins are pale ochre or pale amber and unmarked.

Habitats: Most often found in shallow riffles of warm brooks and streams with swift currents. Occasionally found in lakes or ponds near the mouths of tributary streams.

Natural History: The natural history of the Northern Hog Sucker is similar to that of the White and Longnose Suckers, pp. 99 and 100, respectively, except that the Hog Sucker is nonmigratory. Like the other two species, the Hog Sucker spawns in very shallow riffles with several males attending each ripe female. The eggs are broadcast at random over the gravel bottom; many are almost immediately devoured by dace or chubs.

This species' diet consists of aquatic insects, mollusks, and crustaceans. It sucks its food off the bottom and will use its snout to overturn small rocks in search of food.

Local Names: Black Sucker, Riffle Sucker, Stoneroller, Bighead Sucker, Hognose Sucker

56. Spotted Sucker

Minytrema melanops

Field Marks:

- Body is elongated, heavier ahead of the dorsal fin than behind it.
- Scales are medium-sized.
- Lateral line is incomplete and poorly developed in adults, absent in juveniles.
- Mouth is subterminal.
- Dorsal and ventral fins originate ahead of the midpoint of the total length.
- Anal fin is quite long.
- Caudal fin is moderately forked with triangular lobes.

Adult Size: 6–15 in. (15.24–38.1 cm) average

Color: The upper back, head, and sides are dark brown or olive. The sides, belly, and underparts of the head are whitish or silvery. The sides and back are marked with several horizontal rows of dark spots. All the fins are pale brown or pale amber and are unmarked.

Habitats: The Spotted Sucker is most commonly found in creeks or small rivers with hard bottoms of gravel, sand, or clay. Siltation of many of its native waters has reduced its range.

Natural History: Very little is known of the life of the Spotted Sucker. The males become brightly colored and develop nuptial tubercles in spring and early summer, which suggests that spawning takes place at that time. It is presumed that it spawns in the same manner as other members of this family.

The bulk of the diet of the Spotted Sucker is small aquatic invertebrates.

Local Names: Black Sucker, Striped Sucker

■ BULLHEAD CATFISHES

Order Siluriformes
Family Ictaluridae

Field Marks:

- Body is robust, pot-bellied, and compressed laterally toward the rear.
- No external scales on the body or the head, making the skin appear completely scaleless.
- Lateral line is usually complete and well developed, incomplete in some species.
- Head is flattened with 2 barbels on top near the snout, 4 barbels on the chin, and 1 barbel at each corner of the mouth. The eyes are small.
- Mouth is wide with many small teeth on the jaws and inside the mouth. The snout may slightly overhang the mouth or the jaws may be of equal length.
- Eight fins: paired pectorals set very close behind the gill covers, equipped with a sharp serrated spine on leading edge; paired ventrals, originating near midpoint of the belly; a single long-based anal; a broad, short caudal that may be rounded, squared, or forked; a large fleshy adipose; and a short-based, rounded dorsal with a stout sharp spine on its leading edge.

Habitats: Catfishes are generally found in slow-moving or still waters with muddy or soft bottoms but are not restricted to this type of habitat. They may be encountered in any fresh water, except in very cold habitats.

Natural History: Most catfishes spawn in spring or early summer. Depending on the species, males, females, or both sexes may be involved in building the nest. Both parents guard the nest after spawning and continue to protect the dense school of fry until they are are able to fend for themselves.

All catfishes are bottom feeders, becoming most active after dark. Highly developed sensory cells, most heavily concentrated in the barbels, cover their skin. These cells provide the catfishes with a sense of "taste" that extends over their entire bodies, making them much less dependent on their senses of sight and smell than are day-feeding fishes.

Collecting: Small specimens of catfishes are easy to collect with minnow seines, minnow traps, large dip nets, or small live baits fished on rod and reel. Madtoms, which live in swift waters, are most easily caught with a minnow seine and may also be found inside discarded tin cans lying on stream bottoms.

Handling: Catfishes are not particularly demanding and can be kept in a variety of aquarium ecosystems. As scavengers, they will help keep the tank free of excess organic matter. Some specimens may claim and defend a territory in the aquarium and become aggressive toward any other fishes. To prevent their becoming a nuisance, provide each catfish with its own hiding place. Avoid keeping catfishes with much smaller fishes, since the catfishes may eat them.

Take care when handling catfishes to avoid being stuck by the spines on their pectoral and dorsal fins. (See page 13 for handling instructions.)

57. White Catfish

Ictalurus catus

Field Marks:
· Body is deep, may be pot-bellied, tapering to a body compressed laterally near the tail. The head is flattened on top.
· Lateral line is complete but indistinct.
· Mouth is medium-sized with the snout slightly overhanging lower jaw. The chin barbels are white.
· Anal fin has 19–23 rays.
· Caudal fin is moderately forked with rounded lobes.

Adult Size: 6–18 in. (15.24–45.72 cm) average

Color: Upper body and top of head are light slate-gray or steel-blue. Sides and belly are white or cream. No markings on body or fins. Adipose fin is the same color as the back. Other fins are light ochre or light tan.

Habitats: Usually found in ponds, bayous, or slow-moving streams of fresh or slightly brackish water, occasionally found in streams with moderate currents.

Natural History: This fish is between the size, shape, and behavior of bullheads and madtoms, and larger catfishes such as Blue and Channel cats. A May to June spawner, it often migrates short distances to reach suitable spawning areas. The male prepares a nest in the shelter of a large object such as a log or undercut bank. He then seeks out a female and courts her by sweeping her body with his barbels. The pair mates in a head-to-tail stance; the male wraps his caudal fin around the female's head. Eggs and sperm are released in short bursts until the female is spent. Both parents fan the eggs with their fins to keep them free of silt and debris. When fry hatch they move immediately to the water's surface and begin feeding on plankton. As they grow, they become more omnivorous, finally moving to the bottom and taking whatever food is available. The diet includes aquatic insects and crustaceans, worms, and small fishes.

58. Black Bullhead

Ictalurus melas

Field Marks:
· Body is robust, often pot-bellied, compressed laterally toward the tail.
· Lateral line is complete.
· Head is large.
· Mouth is medium-sized with the upper and lower jaws of equal length. The chin barbels are black.
· Anal fin has 15–19 rays.
· Caudal fin is broad with a rounded trailing edge.

Adult Size: 6–12 in. (15.24–30.48 cm) average

Color: Overall body color is dark gray to black above and cream or light yellow below. There are no distinct markings. The fins are uniform dark gray or gray-brown with cream highlights. The barbels are dark gray to black.

Habitats: The Black Bullhead is commonly found in slow-moving or still water over mud or other soft bottom in lakes, ponds, rivers, and streams. It is tolerant of turbidity, siltation, and moderate amounts of pollution.

Natural History: Spawning begins in May or June when sexually mature adults, led by the females, move into shallow weedy areas. The peak of the spawning season varies widely according to locality and may extend into August. The female prepares the nest by sweeping the bottom with her fins and removing material with her head. The completed nest is usually about as wide as the fish is long. When the nest is finished, a male enters the area and courts the female by butting her and sweeping her body with his barbels. Once she accepts him, the female reciprocates the male's gestures. During mating, the male positions himself beside the female with his tail at her head. Spawning occurs when the male wraps his caudal fin around his mate's head. Eggs and sperm are released several times before spawning is complete. During rest periods between successive spawnings the female carefully fans eggs already in the nest. After spawning both nesting parents continue to fan and guard the eggs, which are held together in a gelatinous mass. A nest commonly contains about 200 eggs, although as many as 5000–6000 eggs have been counted in nests of large females. Eggs usually hatch in 5–8 days. After hatching, the fry form a dense, nearly spherical school and remain close to one of the parent fish at all times. The school and parental protection are maintained until the young fish are about 1 in. (2.54 cm) long.

Young and adult Black Bullheads eat almost any animal matter they find. The bulk of their diet is immature aquatic insects, but they also eat clams, snails, leeches, worms, and small fishes.

Local Names: Hornpout, Horned Pout, Black Catfish, Yellow Belly Bullhead

59. Yellow Bullhead

Ictalurus natalis

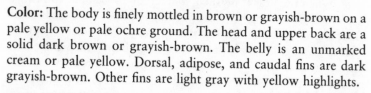

Field Marks:
- Lateral line is complete.
- Head is large.
- Mouth is medium-sized with jaws of equal length. The chin barbels are cream-colored or white.
- Anal fin is long (about 24 rays).
- Caudal fin is long with a rounded trailing edge.

Adult Size: 6–12 in. (15.24–30.48 cm) average

Color: The body is finely mottled in brown or grayish-brown on a pale yellow or pale ochre ground. The head and upper back are a solid dark brown or grayish-brown. The belly is an unmarked cream or pale yellow. Dorsal, adipose, and caudal fins are dark grayish-brown. Other fins are light gray with yellow highlights.

Habitats: The Yellow Bullhead is nearly always associated with dense vegetation in clear, still, or sluggish water in lakes, ponds, and lower portions of streams.

Natural History: When the water temperature reaches 65°F (18°C), mature Yellow Bullheads select nest sites in shallow weedy waters near large rocks, stumps, or overhanging banks. One or both mates construct the nest. A completed nest may be as simple as a small, shallow depression or as elaborate as a 2 ft. (60 cm) deep burrow.

The spawning act is nearly identical to that of the Black Bullhead (p. 106). As many as 1500–4000 eggs are released in gelatinous masses of 300–700 eggs each. Under normal conditions eggs hatch in 5–10 days. Fry form a dense school and maintain close contact with one of their parents. The parents abandon their brood when they reach a length near 2 in. (5.08 cm).

The Yellow Bullhead is omnivorous, browsing along the bottom in search of anything edible.

Local Names: Yellow Catfish, Northern Yellow Bullhead

60. Brown Bullhead
Ictalurus nebulosus

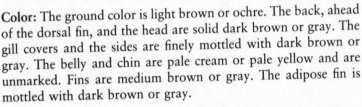

Field Marks:
· Lateral line is complete, nearly straight.
· Mouth is slightly smaller than that of the Black or Yellow Bullhead. The chin barbels are brown or black.
· Anal fin has 20–24 rays.
· Caudal fin is broad with a squared trailing edge.

Adult Size: 6–12 in. (15.24–30.48 cm) average

Color: The ground color is light brown or ochre. The back, ahead of the dorsal fin, and the head are solid dark brown or gray. The gill covers and the sides are finely mottled with dark brown or gray. The belly and chin are pale cream or pale yellow and are unmarked. Fins are medium brown or gray. The adipose fin is mottled with dark brown or gray.

Natural History: The Brown Bullhead begins spawning in May or June when the water temperature reaches 65°F (18°C). Both male and female prepare the nest in water less than 2 ft. (0.61 m) deep over a sand bottom close to some cover such as a log or stump. During spawning, the pair assume a head-to-tail position in which the male embraces the female. Eggs are released in clusters of a few hundred and require several spawnings before they are all released. The parents guard the nest and fan the eggs during their 5–10 day incubation period. Fry form a dense school and remain close to their parents until they are about 2 in. (5.08 cm) long.

Brown Bullheads feed on whatever edible matter they may find.

Local Names: Hornpout, Horned Pout, Marbled Bullhead, Mudcat

61. Stonecat

Noturus flavus

Field Marks:

· Body is slender, elongated, laterally compressed toward the tail.

· Lateral line is incomplete.

· Mouth is smaller than the bullheads', snout slightly over-

hangs the lower jaw. The chin barbels are white.

· Each pectoral fin spine is equipped with a venom gland at its base.

· Adipose fin is fused to the upper leading edge of the caudal fin.

· Caudal fin is long, fan-shaped, with a rounded trailing edge.

Adult Size: 4–6 in. (10.16–15.24 cm) average

Color: The back, head, and sides are light gray or grayish-brown. Belly and chin are white or cream. The dorsal fin is brownish-gray. Other fins are medium gray with cream highlights.

Habitats: Stonecats are found in shallow riffles of small clear streams or rivers with moderate flow.

Natural History: The Stonecat is thought to spawn over an extended period during the late spring and summer. It uses the shallow riffles of streams and rocky shallows of lakes as spawning sites. Eggs are released in a dense gelatinous mass under shelter of a flat rock. One or both parents guard the eggs during incubation and the fry for a short time after their birth.

Stream-dwelling Stonecats grow very slowly, while those living in lakes or ponds tend to grow much faster. They feed mainly on immature aquatic insects that they take from the bottom. In addition to insects, Stonecats will consume small fishes, snails, and other mollusks and crustaceans. Stonecats are inactive during the day, remaining hidden beneath some shelter, usually a rock.

Local Names: Eastern Madtom, Margined Madtom, Stone Catfish, White Cat

62. Tadpole Madtom

Noturus gyrinus

Field Marks:
- Lateral line is incomplete.
- Mouth is small, with jaws of equal length. The chin barbels are reddish-brown.
- Each pectoral fin spine has a venom gland at its base.
- Adipose fin is fused to the back and the top leading edge of the caudal fin.
- Caudal fin is long and fan-shaped, with a rounded trailing edge.

Adult Size: 3–4 in. (7.62–10.16 cm) average

Color: The body, head, and fins are reddish-brown or amber. The belly and lower parts of the gill covers are creamy-yellow. The barbels are the same color as the rest of the head. The fins are usually a slightly lighter shade of the body color.

Habitats: The Tadpole Madtom is found in clear lakes and ponds and in quiet parts of rivers and streams.

Natural History: The Tadpole Madtom is very secretive, spending daylight hours under the cover of rocks or logs. Little is known of its biology. Ripe males and females have been collected from late June through July, suggesting that spawning takes place at that time. Spawning has not been observed, but presumably takes place in the shelter of natural or manmade cavities in the bottom. Both adult fish and egg clusters have been found in empty tin cans.

Like other members of this family, the Tadpole Madtom feeds almost exclusively after dark. It will consume whatever edible matter it finds on the bottom, but the bulk of its diet consists of immature aquatic insects.

Local Names: Tadpole Stonecat

63. Brindled Madtom

Noturus miurus

Field Marks:
- Body is moderately robust—not as deep as the Tadpole Madtom, deeper than the Stonecat.
- Lateral line is incomplete.
- Mouth is small with jaws of equal length. The chin barbels are cream colored.
- Each pectoral fin spine has a venom gland at its base.
- Adipose fin is poorly developed, fused to back and top leading edge of caudal fin.
- Caudal fin is broad, fan-shaped with a rounded trailing edge.

Adult Size: 2–3 in. (5.08–7.62 cm) average

Color: The ground color of the head, body, and fins is a dull reddish-orange. The belly and chin are creamy-yellow. Four black or dark brown saddle shaped marks are on the back and the head. The upper parts of the belly, the lower sides, and the gill covers are spotted with dark brown or gray. The dorsal, adipose, and caudal fins are marked with two dark bands. The anal and ventral fins are each marked with one band.

Habitats: Found in lowland streams with little current over a soft bottom.

Natural History: The few studies that have been done on this species suggest that its biology is similar to that of the Tadpole Madtom. Spawning season is from July to early August. Eggs are often found in discarded tin cans. The male guards the nest after spawning.

Brindled Madtoms feed heavily on algae and detritus that they glean from rocks or other structure.

Local Names: Brindled Stonecat

■ PIRATE PERCHES

Order Percopsiformes
Family Aphredoderidae

Field Marks:
- Body is deep, oval in cross section.
- Scales on the body are medium-sized. There are no scales on the head.
- Lateral line is complete but obscure.
- Head and eyes are large.
- Mouth is medium-sized, with the upper jaw ending at the front edge of the eye.
- Seven fins: broad paired pectorals set just behind the lower part of the gill covers; paired ventrals originating below the pectorals; a single anal fin with 2 weak spines on its leading edge; a broad caudal fin that is slightly indented on its trailing edge; and a single dorsal originating at about midpoint of the back with 2 or 3 weak spines in its leading edge.

Habitats: Found in still or sluggish waters of small weedy lowland rivers, streams, and bogs.

Natural History: The only living species in this family is thought to spawn in the spring. Both parents prepare the nest. Some researchers have suggested that Pirate Perch incubate their eggs under their gill covers, because of the location of the anal opening. In an adult Pirate Perch, the anal opening is located between the lower edges of the gill covers, as in cavefishes, which incubate their eggs under their gill covers.

Collecting: The Pirate Perch may be difficult to find, because of its small size and, if you are searching specifically for it, your efforts may prove fruitless for some time. A minnow seine, large dip net, or minnow trap are suitable devices for catching Pirate Perch. All should be used in densely weeded areas.

Handling: Pirate Perch can be kept in tanks with small sunfishes or perch. The tank should contain one or two sunken branches for the fish to shelter under and should be densely planted.

Pirate Perch can be fed small live or prepared foods.

64. Pirate Perch

Aphredoderus sayanus

Field Marks:
· See family Field Marks.

Adult Size: 3–4 in. (7.62–10.16 cm) average

Color: The head and body ground color is dark ochre; the scales are a contrasting olive or gray. The head is marked with fine mottling in dark brown. The dorsal and caudal fins are medium olive. There is a dark band at the base of the dorsal fin, and a dark stripe at the base of the caudal fin. The anal and ventral fins are light ochre and unmarked. The pectoral fins are transparent, with light ochre at their bases.

Habitats: See family Habitats.

Natural History: See family Natural History.

■ TROUT-PERCHES

Order Percopsiformes
Family Percopsidae

Field Marks:
· Body is slender, elongated with a moderate hump on the back, which peaks at the origin of the dorsal fin.
· Scales on the body are medium-sized. The head is not scaled.
· Lateral line is incomplete, not well defined.
· Head is medium-sized with large eyes.
· Mouth is medium-sized, terminal with the snout very slightly overhanging the lower jaw.
· Eight fins: paired pectorals set low on the body just behind the gill covers; paired ventrals set far forward, originating below the leading edge of the dorsal fin; a single squarish anal fin set far forward on the back; a broad, moderately forked caudal fin

1. Longnose Gar

2. Shortnose Gar

3. Bowfin

4. American Eel

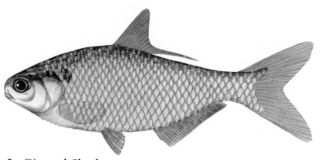

5. Gizzard Shad

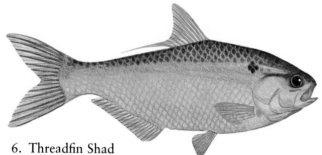

6. Threadfin Shad

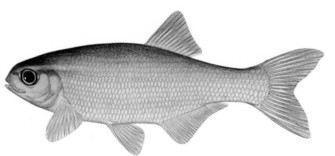

7. Mooneye

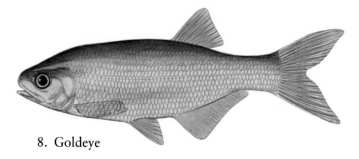

8. Goldeye

9. Golden Trout

10. Cutthroat Trout

11. Rainbow Trout

12. Brown Trout

13. Brook Trout

14. Central Mudminnow

15. Eastern Mudminnow

16. Redfin Pickerel

17. Grass Pickerel

18. Northern Pike

19A. Chain Pickerel
(Juvenile)

19B. Chain Pickerel
(Adult)

20. Mexican Tetra

21. Stoneroller

22. Goldfish

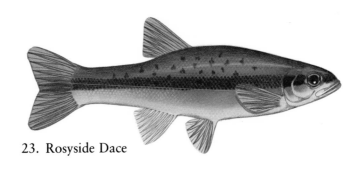

23. Rosyside Dace

24. Lake Chub

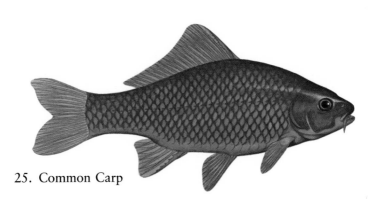

25. Common Carp

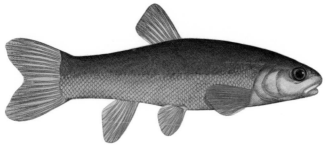

26. Cutlips Minnow

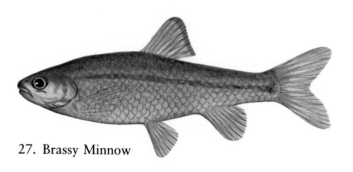

27. Brassy Minnow

28. Central Silvery Minnow

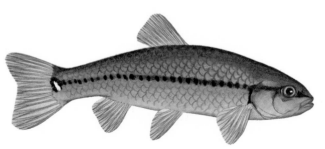

29. Hornyhead Chub

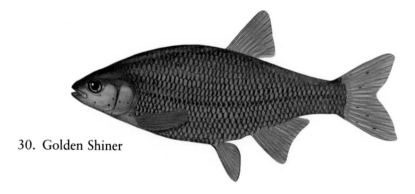

30. Golden Shiner

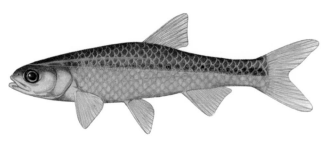

31. Emerald Shiner

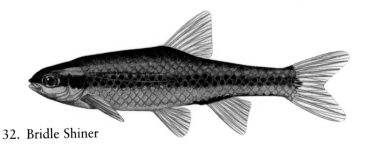

32. Bridle Shiner

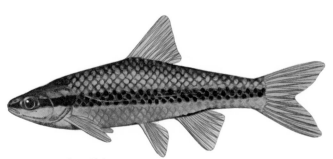

33. Ironcolor Shiner

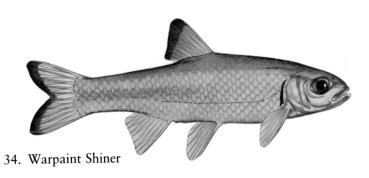

34. Warpaint Shiner

35. Common Shiner

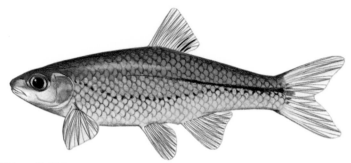

36. Whitetail Shiner

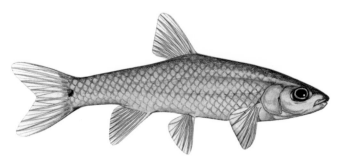

37. Spottail Shiner

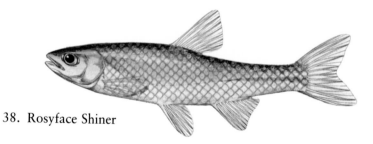

38. Rosyface Shiner

39. Spotfin Shiner

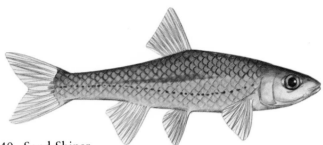

40. Sand Shiner

41. Mimic Shiner

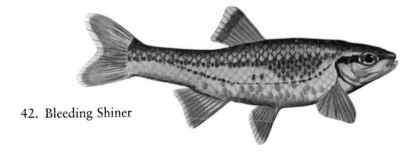

42. Bleeding Shiner

43. Northern Redbelly Dace

44. Southern Redbelly Dace

45. Finescale Dace

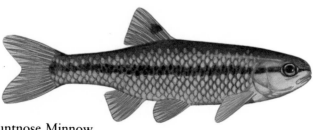

46. Bluntnose Minnow

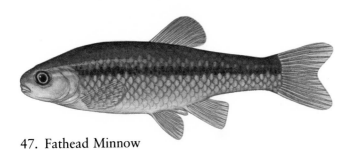

47. Fathead Minnow

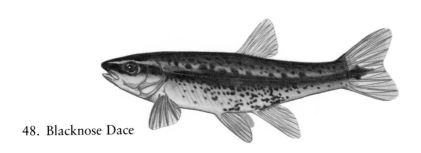

48. Blacknose Dace

49. Longnose Dace

50. Creek Chub

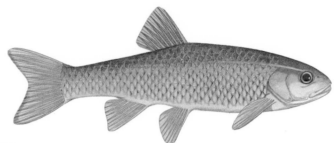

51. Fallfish

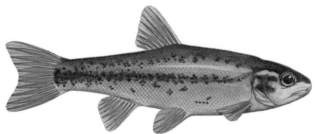

52. Pearl Dace

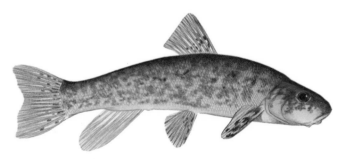

53. Longnose Sucker

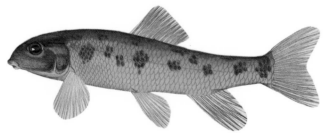

54. White Sucker

55. Northern Hog Sucker

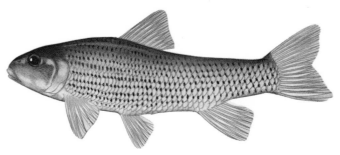

56. Spotted Sucker

57. White Catfish

58. Black Bullhead

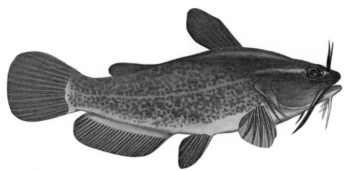

59. Yellow Bullhead

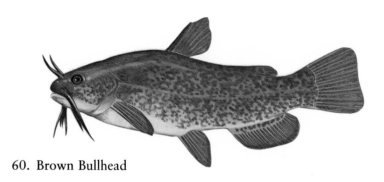

60. Brown Bullhead

61. Stonecat

62. Tadpole Madtom

63. Brindled Madtom

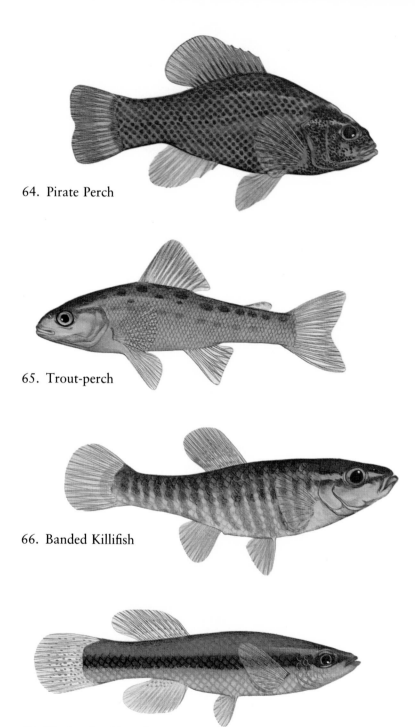

64. Pirate Perch

65. Trout-perch

66. Banded Killifish

67. Blackstripe Topminnow

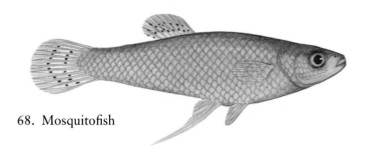

68. Mosquitofish

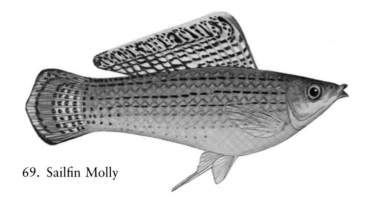

69. Sailfin Molly

70. Brook Silverside

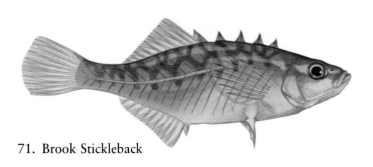

71. Brook Stickleback

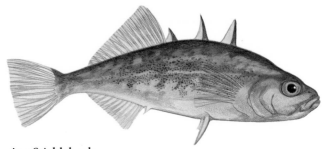

72. Threespine Stickleback

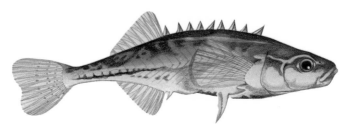

73. Ninespine Stickleback

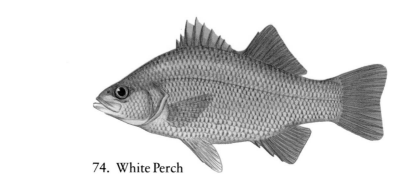

74. White Perch

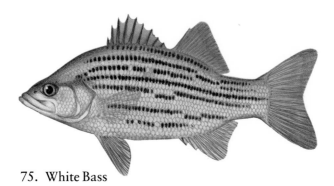

75. White Bass

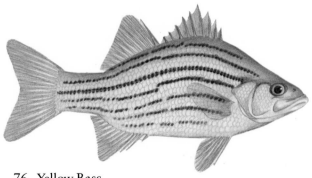

76. Yellow Bass

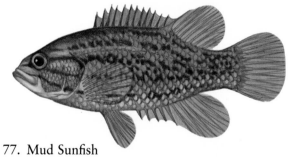

77. Mud Sunfish

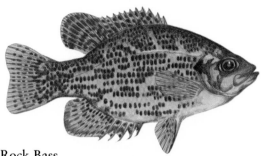

78. Rock Bass

79. Flier

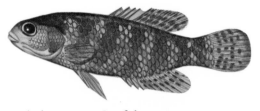

80. Everglades Pygmy Sunfish

81. Banded Pygmy Sunfish

82. Blackbanded Sunfish

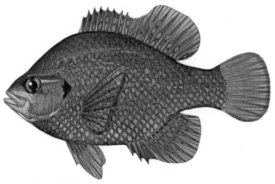

83. Bluespotted Sunfish

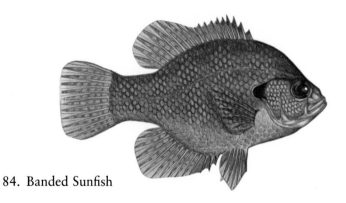

84. Banded Sunfish

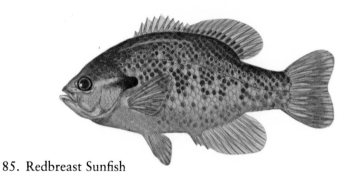

85. Redbreast Sunfish

86. Green Sunfish

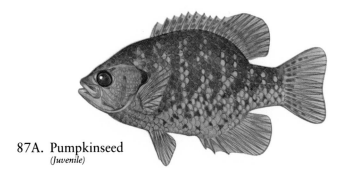

87A. Pumpkinseed
(Juvenile)

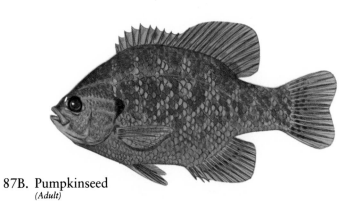

87B. Pumpkinseed
(Adult)

88. Warmouth

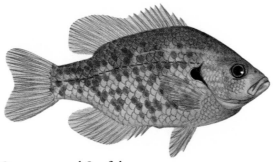

89A. Orangespotted Sunfish
(Nonbreeding adult)

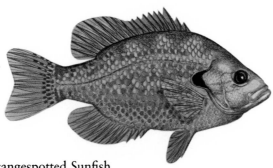

89B. Orangespotted Sunfish
(Adult — breeding male)

90A. Bluegill
(Juvenile)

90B. Bluegill
(Adult)

91. Longear Sunfish

92. Redear Sunfish

93. Spotted Sunfish

94.
Bantam Sunfish

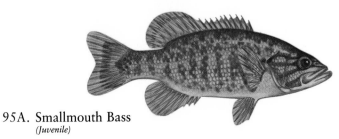

95A. Smallmouth Bass
(Juvenile)

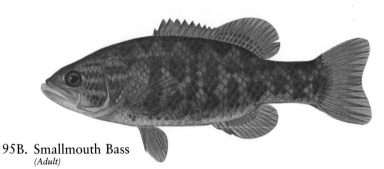

95B. Smallmouth Bass
(Adult)

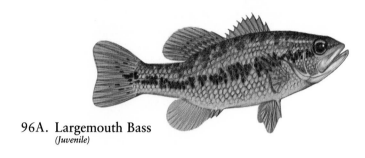

96A. Largemouth Bass
(Juvenile)

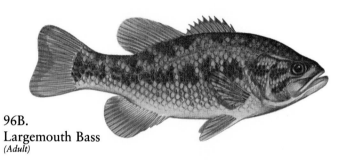

**96B.
Largemouth Bass**
(Adult)

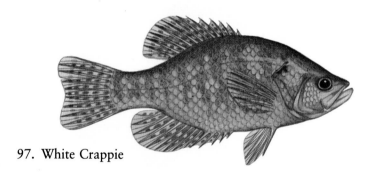

97. White Crappie

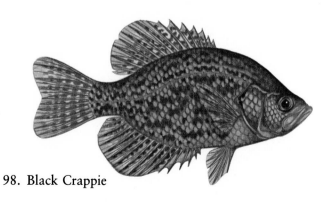

98. Black Crappie

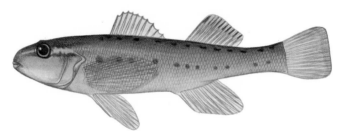

99. Eastern Sand Darter

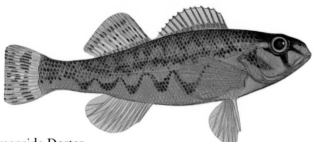

100. Greenside Darter

101. Rainbow Darter

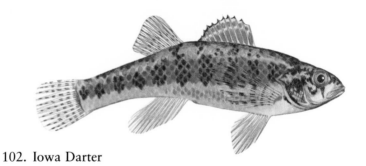

102. Iowa Darter

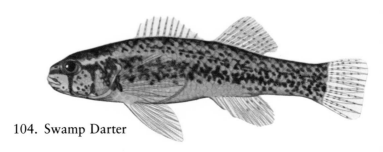

103. Fantail Darter

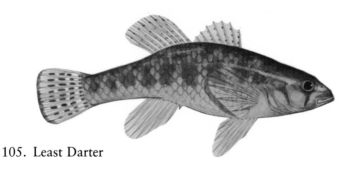

104. Swamp Darter

105. Least Darter

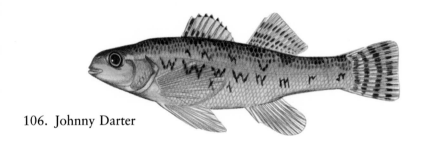

106. Johnny Darter

107. Tessellated Darter

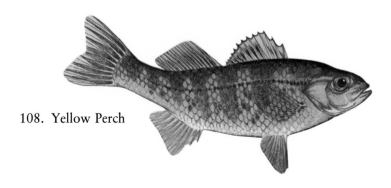

108. Yellow Perch

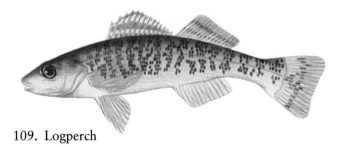

109. Logperch

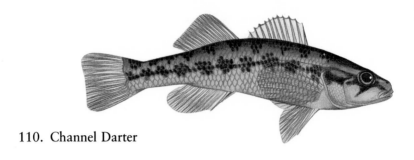

110. Channel Darter

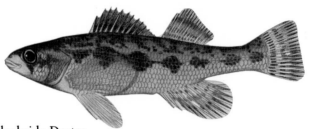

111. Blackside Darter

112. River Darter

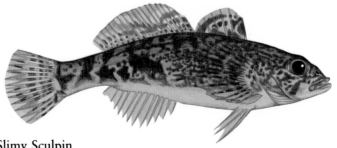

113. Slimy Sculpin

with triangular lobes; a small fleshy adipose fin just ahead of the caudal fin; and a single broad-based dorsal fin orginating unusually far forward on the back.

Habitats: Trout-Perch can be found in large clear lakes except when spawning in shallow turbid streams.

Natural History: Collections of ripe Trout-Perch specimens suggest that it begins spawning in May and may continue well into August. It migrates short distances into turbid shallow streams to spawn or may use sand bottoms in shoal areas of lakes if no streams are available. The incubation period is 20 days.

Trout-Perch feed on immature aquatic insects, small crustaceans, and small fishes. It feeds almost exclusively after dark, preferring to hide in brush or vegetation during the day.

This is one of only 2 living species that represent this family.

Collecting: Collecting for this species should be done at night with a minnow seine or minnow trap. When the fish are spawning in shallow streams, they can be caught in minnow seines or large dip nets.

Handling: The Trout-Perch should be kept over a sand or fine gravel bottom. The tank may be decorated with several large rocks and perhaps a few plants to provide shelter. It can be fed small live foods such as brine shrimp or mosquito larvae and can be conditioned to accept dried foods.

65. Trout-Perch

Percopsis omiscomaycus

Field Marks:
· See family Field Marks.

Adult Size: 3–4 in. (7.62–10.16 cm) average

Color: The body of the Trout-Perch appears translucent. The

body and head are pale straw or flesh colored. There is a horizontal row of 9–12 dark brown spots along the upper back and a similar row of 7–12 lighter spots along the sides, following the curve of the lateral line. All fins except the adipose are very pale yellow. The adipose fin is slightly darker and faintly marked with a few small brown spots along its upper edge.

Habitats: See family Habitats.

Natural History: See family Natural History.

■ KILLIFISHES

Order Atheriniformes
Family Cyprinodonidae

Field Marks:
· Body is elongated, robust, slightly compressed laterally, oval in cross section.
· Scales are medium- to large-sized. Some species have scales on the gill covers and cheeks; others have no scales on the head.
· Lateral line is absent from the body, but a few lateral line pores appear on the head.
· Head is broad across the top. Eyes are small- to large-sized.
· Mouth is small- to medium-sized and tilted upward; the lower jaw projects beyond the upper jaw.
· Seven fins: broad paired pectorals located just behind the gill covers, about one third of the way up the body; paired ventrals (absent in some species) that originate at or slightly ahead of the midpoint of the belly; a single, short-based anal usually set far forward; a broad fan-shaped caudal with rounded or squared trailing edge; and a single rounded dorsal originating slightly behind the midpoint of the back. All fins are soft-rayed and have no spines.

Habitats: The killifishes usually occur in shallow, weedy areas of fresh or brackish water. They are most common in warm-water habitats. Some species are marine.

Natural History: The killifishes generally spawn in the spring or summer. They build no nests. Their mouths are evolved for feeding at the water's surface; they feed mainly on adult aquatic or terrestrial insects.

Collecting: Minnow seines, flat dip nets hung on poles, standard large dip nets, minnow traps, tiny live baits on barbless hooks, or very small artificial flies on barbless hooks are all appropriate methods for collecting killifishes. The members of this family are often locally abundant and are easily caught.

Handling: Killifishes are easy to keep in an aquarium. Their only special need is that they be given foods that will remain floating on the surface; flake-style dry foods are probably the most convenient.

Killifishes are easy to breed in captivity and are avidly sought as one of the most popular native aquarium fishes.

66. Banded Killifish

Fundulus diaphanus

Field Marks:
· Body is robust anteriorly, compressed laterally toward the tail.
· Scales are medium-sized.
· Lateral line is absent, except for a few pores on the head.
· Eyes are large.
· Caudal fin is relatively small, fan-shaped, rounded on its trailing edge.

Adult Size: 2–3 in. (5.08–7.62 cm) average

Color: The back and upper parts of the head are dark olive or dark ochre. The sides and lower parts of the head are silvery-white. The sides are marked by a series of 14–22 olive vertical bars; these are transversed by a darker bluish lateral stripe. The pectoral and ventral fins are nearly transparent and may be flushed with pale ochre. The anal fin has several rows of pale ochre spots radiating outward from its base. The caudal fin has a narrow olive or light brown band along its base. The dorsal fin is marked with irregular rows of olive or brown spots. The ground color of all fins is a very pale tan.

Habitats; In fresh water the Banded Killifish is found in quiet shallows of warm lakes, ponds, and slow-moving streams over sand bottom away from dense vegetation. It also occurs in brackish pools and streams in salt marshes.

Natural History: The Banded Killifish usually begins spawning in May or June and may continue into August. As the spawning season approaches the males enter shallow water near dense vegetation, where they claim small nesting territories. The males become aggressive toward one another; butting and nipping but seldom doing any damage. During the spawning season, the males take on an intense bluish-green color, while the females remain comparatively pale. Courtship.involves active pursuit of a chosen mate by the male; during this chase the female's genitals become enlarged in preparation for spawning. When she is ready to mate she releases a single egg; which remains attached to her body by a fine thread. This apparently triggers the spawning act, and the male pursues her even more vigorously, forcing her into dense vegetation, where he presses close to her side. As small clusters of eggs are released, the male vibrates violently, releasing his sperm and fertilizing the eggs. The eggs are held together by a fine thread, but separate from each other as they sink into the vegetation. The average female will produce 50 eggs. The incubation period lasts about 10–12 days at 70°–80°F (21°–27°C).

Banded Killifish feed mainly on adult aquatic insects, but they will occasionally take almost any available food at any level in the water column.

Local Names: Freshwater Mummichog, Freshwater Killy, Topminnow. Tommycod, Tomcod, Barred Minnow

67. Blackstripe Topminnow
Fundulus notatus

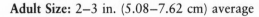

Field Marks:
- Body is elongated, more slender than in the Banded Killifish.
- Scales are medium-sized. Cheek and gill covers are scaled.
- Lateral line is absent.
- Head is smaller than in most killifishes. The eyes are large.
- Anal and dorsal fins originate far back on the body.
- Caudal fin is fan-shaped, rounded on its trailing edge.

Adult Size: 2–3 in. (5.08–7.62 cm) average

Color: The back and upper parts of the sides are reddish- or yellowish-brown. A distinct broad, black or purplish lateral band begins behind the upper jaw and ends at the rear edge of the caudal peduncle. The lower portions of the sides, lower portions of the head, and the belly are silvery-white. The fins are very pale tan. The dorsal, caudal, and anal fins are marked with several rows of tiny brown spots.

Habitats: The Blackstripe Topminnow is usually found in shallow water in lakes, ponds, and slow-moving rivers or streams. It is usually encountered in large schools near the surface.

Natural History: The spawning season of this species continues throughout the summer months. During the season, spawning pairs separate from the larger school and claim a territory near shore in shallow water. Both fish will defend this territory against intruders. Spawning takes place with the male and female lying side-by-side while they vibrate together and one egg is released and fertilized. Twenty to thirty eggs are released in this manner, with the male (occasionally the female) ending each release with a flip of the tail that throws the egg into nearby vegetation.

The Blackstripe Topminnow is most active in the morning and again in the late afternoon to evening. It feeds mainly on terrestrial insects but will also eat aquatic insects, crustaceans, snails, and algae.

■ LIVEBEARERS

Order Atheriniformes
Family Poeciliidae

Field Marks:
- Body is generally robust, often pot-bellied, generally oval in cross-section, broad across the back of the head, and laterally compressed toward the tail.
- Scales are large. Most species have no scales on the head.
- Lateral line is usually absent (a few sensory pores appear on the head), but may be present and incomplete or complete.
- Head is roughly triangular in profile, with a rounded or pointed snout and usually large eyes.
- Mouth is small, terminal, tilted upward with the lower jaw projecting beyond the upper.
- Seven fins: paired pectorals set a short distance behind the gill covers; paired ventrals originating ahead of the midpoint of the belly; well forward on the body is a single anal fin, whose leading edge may be developed into a long rodlike projection; a broad, usually fan-shaped caudal fin; and a single, usually short-based rounded or squared (the Sailfin Molly has a long-based dorsal) dorsal fin. The fins have no spines.

Habitats: Members of this family are most often found in warm shallow waters with dense vegetation.

Natural History: The major difference between these fishes and the topminnows and killifishes is that the eggs are fertilized before they are released by the female. The male's anal fin is developed into a long rod for the purpose of implanting the sperm. (See details in species accounts.)

Collecting: A minnow seine, fished through weedy areas where these species are suspected to live, is the best way to collect them.

Handling: The two livebearers included here are among the very few native North American fishes to have been given much attention as aquarium fishes. They are well suited to life in the home aquarium, being easy to maintain in a thickly planted tank and generally easy to breed. They prefer live food, but will readily accept dried food; flake foods that remain in the surface film for extended periods are best for these fishes.

68. Mosquitofish
Gambusia affinis

Field Marks:
- Body is stout, slightly compressed laterally toward the tail.
- Scales are large. There are large scales on the gill covers and smaller scales on the cheeks.
- Lateral line is absent. There is a sensory canal on each side of the head.
- Head is small with a small, upturned terminal mouth. Eyes are large.
- Dorsal fin is small, rounded. Caudal fin is broad, fan-shaped. Anal fin, in males, is developed into a long, thin rod.

Adult Size: males, 1 in. (2.54 cm); females, 2 in. (5.08 cm) average

Color: The entire fish—body, head, and fins—is a very pale whitish-olive. The belly may be white. The dorsal and caudal fins are marked with 2 rows of small dark spots.

Habitats: The Mosquitofish prefers warm sluggish streams, where it will be found in backwaters and oxbows in and around dense vegetation.

Natural History: The Mosquitofish's spawning act, and the special reproductive system evolved for it, are very unusual. Mosquitofish spawn over a 10–15-week period that includes most of the summer. The males court females throughout this period and a female may give birth to several broods. The eggs are fertilized as the mating pair lie side-by-side while the male uses specially developed muscles to turn his anal fin toward his mate and insert it in her genital canal. The sperm is moved through a canal in the elongated third ray of the anal fins. The living sperm is held in reserve in a specially evolved pouch in the female's belly.

Several batches of eggs may be fertilized by the sperm from one mating. The number of young in each brood varies widely from a few to several hundreds. The eggs require nearly a month of incubation before the young emerge from their mother. The young grow rapidly and may be sexually mature in their first summer. Most individuals die after their first spawning; only a few fish survive through their second summer.

The diet of the Mosquitofish is mainly insects, snails, and crustaceans. It also consumes plants, such as duckweed, but may do so accidentally.

Local Names: Potgut Minnow

69. Sailfin Molly
Poecilia latipinna

Field Marks:
· Body is very stout, tapering very little from the back of the head to the caudal fin. It is slightly pot-bellied.
· Scales are large. The body scales extend onto the top of the head. No scales on the cheeks or gill covers.
· Lateral line is absent.
· Head is small with large eyes.
· Mouth is small, terminal, with small upturned jaws that extend beyond the tip of the snout.
· In males, the third ray of the anal fin is developed into a long rodlike extension. The long-based dorsal fin is unusually high, square along its top edge. The caudal fin is broad, triangular along its trailing edge.

Adult Size: males, 3 in. (7.62 cm); females, 4 in. (10.16 cm) maximum

Color: The back and upper parts of the sides and head are pale olive-green. The belly and chin are white or cream. Several thin dark lateral stripes run along the sides. The pectoral, ventral, and anal fins are white or transparent. The base color of the caudal fin is white or pale tan; its rear edge is marked by a thin black margin. There are 5 vertical rows of orange spots surrounded by black borders radiating out from the base of the caudal fin; some of these spots extend onto the caudal peduncle. There is a bright orange crescent on the caudal fin. The dorsal fin has a white ground marked with several rows of black, wavy, vertical lines. The leading edge of the dorsal fin is dusky or black and the top edge is bright yellow.

Habitats: Found in a wide range of warm-water habitats, either in fresh, brackish, or salt water. The Sailfin Molly may be encountered in lakes and ponds, springs, slow-moving rivers and streams, drainage ditches, and pools or streams in salt marshes.

Natural History: The natural history of this species is identical to that of the Mosquitofish (p. 121).

■ SILVERSIDES

Order Atheriniformes
Family Atherinidae

Field Marks:
· Body is slender, elongated, appears translucent.
· Scales are small; the cheeks and gill covers are scaled.
· Lateral line is absent. There are sensory pores on the head.
· Head is small, distinctly pointed. The eyes are large.
· Mouth is small, tilted upward, with jaws of equal length.
· Eight fins: large paired pectorals set at about the midpoint of the sides, just behind the rear edge of the gill covers; small paired ventrals originating ahead of the midpoint of the belly; a long-based anal fin; a moderately or deeply forked caudal fin with triangular lobes; and 2 separate dorsal fins. The first dorsal fin is short and triangular, with supporting spines; the second is soft-rayed and originates far back on the body.

Habitats: The silversides are widely distributed throughout the world, although their freshwater range is not extensive. In fresh water they are found in clear lakes and streams.

Natural History: The freshwater silversides are generally spring or early summer spawners. Silversides grow very rapidly, feeding on a diet of tiny aquatic and terrestrial insects.

Collecting: A minnow seine, fished in open water, is probably the best method of catching silversides. They can also be taken in a large dip net.

Handling: Several silversides should be kept together, as they are schooling fish. They should be kept in a tank with a sand or gravel bottom with densely planted areas.

Since silversides are surface feeders, they should be offered foods that will remain suspended in the surface film for an extended period; flake-style dry food is probably the most convenient.

70. Brook Silverside

Labidesthes sicculus

Field Marks:
· See family Field Marks, p. 123.

Adult Size: 3–4 in. (7.62–10.16 cm) average

Color: The ground color varies from olive to tan to ochre, depending upon the colors of the environment from which the fish is taken. There are two narrow, dusky or black lateral stripes, one above and one below a slightly broader, light-colored stripe. The lower sides and belly are translucent silvery-white. The fins, unmarked, are transparent or extremely pale olive or tan.

Habitats: The Brook Silverside is usually found in large clear lakes or slow-moving streams with dense vegetation over sand bottom.

Natural History: Brook Silversides spawn from May through August. For several days before mating, one to three males will pursue a single female. Only one male will actually mate with the pursued female. The eggs and sperm are released as the mating pair glides at an oblique angle downward from the surface of the water. There is a long threadlike filament attached to each egg; this filament adheres to whatever it touches as the egg sinks to the bottom. The incubation period is 8 days at a water temperature of 77°F (24°C).

Brook Silversides spend nearly all their time within a few inches of the surface. They spend their daylight hours darting about and often leaping several inches out of the water; they continue this behavior on moonlit nights, but lie still at the surface on dark nights. Silversides are attracted by bright light, so fishing a seine or dip net for them at night with the aid of a flashlight (where it is legal) is an effective collecting technique.

Young silversides feed on plankton; the adults feed mainly on adult insects, which they take at the water's surface.

Local Names: Skipjack

■ STICKLEBACKS

Order Gasterosteiformes
Family Gasterosteidae

Field Marks:
- All sticklebacks are small fishes with laterally compressed bodies and very slender caudal peduncles.
- There are no scales on the body or head; in some species the body is partially covered with bony plates.
- Lateral line may be complete or incomplete, but is not well developed; in some species there is a row of bony scutes along the lateral line.
- Head is medium-sized with large eyes.
- Mouth is small to medium-sized and tilted upward; there are small, well-developed teeth on the jaws.
- Eight fins: paired soft-rayed pectorals located just behind the gill covers, about midway on the sides; paired ventrals that originate just below the pectorals and are modified into sharp spines with a single attached membrane; a single long-based soft-rayed anal fin with one short spine on its leading edge; a long, generally fan-shaped soft-rayed caudal fin, which may be

squared, rounded, or slightly forked along its trailing edge; the dorsal fin is composed of 2 distinct fins, the first having 3–12 short spines with single attached membranes, and the second, which is long-based, has one short spine on its leading edge, and is generally square-shaped or triangular; the second dorsal is soft-rayed behind the spine.

Habitats: The sticklebacks are usually found in shallow water near the shores of streams, ponds, or estuaries, usually near or sheltering in dense vegetation.

Natural History: The sticklebacks are some of the most interesting fishes available to the fish watcher. They exhibit elaborate prespawning and mating behavior in which the male builds an intricate nest suspended above the bottom, usually on the stem of a plant. Sticklebacks breed readily in captivity, making them ideally suited for the home aquarium.

Their natural foods include immature and adult aquatic and terrestrial insects, small aquatic worms, crustaceans, and mollusks.

Collecting: A minnow seine, dip net, or minnow trap are the best tools for collecting sticklebacks. Once sticklebacks have been located, they should not be difficult to catch.

Handling: The sticklebacks are generally undemanding in the aquarium. Unfortunately, the males often become aggressive toward one another in captivity, so it is best to keep only one pair of sticklebacks in a tank, allowing them plenty of room. They can be fed a broad range of live or prepared foods; in short, whatever you find most convenient.

71. Brook Stickleback

Culaea inconstans

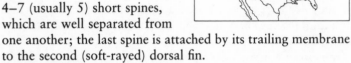

Field Marks:

· Lateral line is incomplete; its pores are sheathed in tiny bony plates.
· First dorsal fin is composed of 4–7 (usually 5) short spines, which are well separated from one another; the last spine is attached by its trailing membrane to the second (soft-rayed) dorsal fin.
· Pectoral fins are rounded.
· Caudal fin is rounded along its trailing edge.

Adult Size: 2 in. (5.08 cm) average

Color: The ground color on the body is medium olive-green; this becomes very dark, nearly black, in breeding males. The body is marked in one of two ways: either with a distinct pattern of wavy, Y-shaped, dark lines along the back and upper sides, or with light, irregularly shaped spots over the entire body, except on the belly. The belly is always white or cream. The first dorsal fin (spined) is olive. The second dorsal (soft-rayed), caudal, anal, and ventral fins are pale tan or cream. The pectoral fins are transparent, sometimes flushed with pale tan or white.

Habitats: In fresh water, the Brook Stickleback is found in the quiet weedy areas of clear, cold streams or small spring-fed ponds. It is also commonly found in shallow pools in salt marshes.

Natural History: The Brook Stickleback begins spawning in late April in the southern part of its range and as late as July in the north. At the beginning of the spawning season males move into shallow water near shore and claim small territories. Each fish selects a nest site at or near the base of a stalk of grass or reed. Two nests are usually built during each spawning season, with the first constructed of small pieces of grass and the latter made of filamentous algae. Each nest is attached to the stalk of a rooted plant; nests are roughly spherical and have a hole in one end. When a nest is finished, the male entices a mate to its entrance by displaying his intense breeding colors and erecting his fins. When the female is near the nest's opening, the male drives her into the nest by nipping and prodding her from behind. Immediately after the female deposits her eggs, the male drives her from the nest, forcing her to break another hole. The male then enters the nest and fertilizes the eggs; he remains in the nest with only his head protruding from one of the openings. He aerates the eggs by fanning them with his pectoral fins, increasing the intensity of the fanning as the eggs continue to develop. The male remains in the nest until the eggs have hatched. Both males and females return to somewhat deeper water at the completion of the nesting period.

Immature aquatic insects, mollusks, and crustaceans make up the diet of the Brook Stickleback.

Local Names: Common Stickleback, Pinfish, Variable Stickleback, Fivespine Stickleback, Sixspine Stickleback

72. Threespine Stickleback
Gasterosteus aculeatus

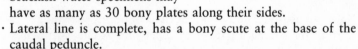

Field Marks:
· Typical of family (p. 125) except as follows:
· Freshwater specimens have no scales and usually no bony plates on the sides. Salt- and brackish-water specimens may have as many as 30 bony plates along their sides.
· Lateral line is complete, has a bony scute at the base of the caudal peduncle.
· Trailing edges of the pectoral fins are straight.
· First dorsal fin consists of 3 (occasionally 4) stout spines (the first 2 are taller than the third). Each spine has a single attached membrane.
· Caudal fin is squared or slightly convex along its trailing edge.

Adult Size: 2 in. (5.08 cm) average

Color: The back and upper sides are dark olive-green to silvery-gray. Below this the sides are metallic silver with blue, green, or brown reflective surfaces.

The fins of nonbreeding specimens are pale tan or amber. The dorsal fin spines are dark olive-green or gray, with pale tan or transparent trailing membranes.

During the spawning season males become flushed with intense red or orange and have bright blue eyes, while females have bright pink on their bellies. The fins of breeding specimens are flushed with pale red, pink, or orange.

Habitats: The Threespine Stickleback is found in small ponds and streams as well as in brackish salt marsh pools.

Natural History: The spawning ritual of this species is nearly identical to that of the Brook Stickleback (p. 127). It differs from the Brook Stickleback in that it builds its nest directly on the bottom. The nest site is usually in shallow water over a sand bottom. The nest is made of small twigs and other vegetable matter that the male cements together with a mucus-like substance secreted from his kidneys.

Threespine Sticklebacks feed on aquatic insects, both immature and adult, crustaceans, worms, fish eggs, and fry (including the fry of its own species).

Local Names: Twospine Stickleback, Common Stickleback, Banstickle, Saw-finned Stickleback, Eastern Stickleback, European Stickleback, Pinfish

73. Ninespine Stickleback
Pungitius pungitius

Field Marks:
· Body is laterally compressed, is not scaled, and has no bony plates.
· Lateral line is complete but inconspicuous; the rear parts of the lateral line may be covered with small bony scutes.

- Pectoral fins are long with slightly rounded trailing edges.
- First dorsal fin has 8–11 (usually 9) short spines, with single trailing membranes; second dorsal fin has one short spine ahead of its soft-rayed portion.
- Caudal fin is fan-shaped, slightly forked, with rounded lobes.

Adult Size: 2.5 in. (6.25 cm) average

Color: The ground color of the upper back and sides (to the lateral line) is light brown or light olive. Below the lateral line, the sides and belly are very pale tan or cream. The upper back is marked with indistinct dark saddle-shaped spots and smaller irregular spots. Irregular spots may occur below the lateral line. Often, a dark brown or black patch begins below the gill covers and extends along the underbelly to the origin of the anal fin. The first dorsal fin spines are dark brown, with lighter membranes. The other fins are a very pale transparent tan. The caudal fin may be marked with several rows of tiny, dark brown spots.

Habitats: The Ninespine Stickleback is most often found in small clear streams, small ponds, and brackish salt marsh pools.

Natural History: During the summer spawning season, both sexes of the Ninespine Stickleback are brightly colored and very aggressive toward other fishes. The male builds a small tunnel-shaped nest in typical stickleback fashion, cementing the nest together with a secretion from his kidneys and leaving a hole in one end of the completed nest. He then entices a mate to enter the nest by displaying his erected fins and his spawning colors. The female enters the nest and deposits 20–30 eggs there; she is followed immediately by the male, who drives her out and then fertilizes the eggs. The male may then entice several more females to enter the nest (one at a time) and repeat the spawning process. He then enters the nest and remains there, guarding the eggs until they have hatched and then the fry, which remain with him in the nest until they are about 2 weeks old. After the first brood has left the nest, he may construct another nest in preparation for additional spawnings.

The diet of the Ninespine Stickleback is typical of the family and includes a high percentage of aquatic invertebrates.

Local Names: Pinfish, Tenspine Stickleback

■ TEMPERATE BASSES

Order Perciformes
Family Percichthyidae

Field Marks:
· Body is elongated, moderately deep, and moderately compressed laterally in all species.
· Body and head are covered with medium-size ctenoid scales.
· Lateral line is complete and conspicuous.
· Head is medium to large, with many small teeth on the jaws and few in the mouth itself. Upper jaw ends at or before the midpoint of the eye.
· Eight fins: paired pectorals originating just behind the trailing edge of the gill covers, slightly below the midline of the body; paired ventrals, with 1 stout spine on their leading edges, originating below and slightly behind the origin of the pectorals; a single, short-based, squared or triangular ventral originating behind the midpoint of the belly with 3 spines on its leading edge; a broad, moderately or slightly forked caudal fin; and 2 dorsal fins, which are either slightly separated or joined by a very short membrane.

Habitats: Freshwater species of this family are typically found in dense schools in large lakes or rivers.

Natural History: The White and Yellow Basses live their lives entirely in fresh water, while White Perch can also be anadromous, living in salt water but returning to fresh water to spawn. All three species enter small tributary streams or rivers in vast numbers just prior to the early spring spawning season. No nest is prepared, and spawning is accomplished at or near the surface of the water by a ripe female typically accompanied by several males. The adhesive eggs usually hatch in the gravel a short time after the completion of spawning. Growth rates vary widely according to environmental conditions but are usually quite rapid, with females growing faster than males.

The diet consists of plankton at first, followed by aquatic insects or crustaceans, and later by small fishes.

Collecting: The best method for collecting the young of these fishes (in sizes suitable for the aquarium) is to seine-net a tributary stream where the adults are known to spawn. This should be done a month or two after the completion of spawning. Aquarium-size specimens are not usually found in larger bodies of water.

Handling: These are hardy fishes and should require only the usual attention to clean water and appropriate foods.

74. White Perch

Morone americana

Field Marks:
· Body is moderately deep, but less deep than in the White or Yellow Bass, and usually less deep in adults than juveniles.
· Scales are medium-sized.
· Lateral line is complete and very distinct.
· Mouth is smaller than in the other species in the family, with upper jaw ending at the front margin of the eye.
· Dorsal fins are connected by a conspicuous membrane.

Adult Size: 8–10 in. (20.32–25.4 cm) average

Color: Upper back and most of the sides are bronze, golden, or silvery, depending on the color of the water from which the fish is taken. Lower sides and belly are white or cream. Ventral and anal fins are pale gray or white. Caudal fin ranges from amber to gray. Dorsal and pectoral fins are amber to pale brown.

Habitats: Landlocked populations of White Perch are most common in ponds and lakes with numerous tributary streams. They thrive in warm or cold water, near or far from shore, over any type of bottom.

Natural History: Typical of its family, the White Perch is a spring spawner, beginning in May and continuing through mid- or late June. Where it has access to tributary streams, it will ascend these in compact schools. Spawning takes place when large groups of fish gather in shallow water and randomly release the eggs and sperm, which are left to sink to the bottom. The fertilized eggs are adhesive. An average female may produce as many as 250,000 eggs. The young fish grow rapidly on a diet of plankton and then aquatic insects. The adults feed on a wide range of foods, including insects and small fish.

Local Names: Silver Perch

75. White Bass
Morone chrysops

Field Marks:
· Body is moderately deep and laterally compressed; slightly deeper than in the White Perch, not so deep as the Yellow Bass.
· Scales are medium-sized.
· Lateral line is complete, distinct.
· Mouth is medium-sized, with the upper jaw ending at about midpoint of the eye.
· Dorsal fins are separate.
· Spines on the anal fin are evenly graduated.

Adult Size: 9–15 in. (22.86–38.10 cm) average

Color: Sides and belly are bright silvery-white. Upper back is light tan or light olive. Pectoral, ventral, and anal fins are white or very pale gray. Caudal and dorsal fins are very pale tan or gray. The sides are marked with 5–7 dark gray or black horizontal stripes.

Habitats: Usually found in dense schools near the surface of large ponds, lakes, and rivers, where it feeds on small fishes.

Natural History: As the early spring spawning season approaches, mature males move into tributaries followed, as much as a month later, by the females. As the females arrive on the spawning grounds, the schools remain separated according to sex, with the females holding in deeper water. To begin the spawning act, a female will rise toward the surface as she approaches the schooling males. This rise will attract several males, which will crowd around her as the eggs and sperm are simultaneously released. The eggs are adhesive and attach themselves to the gravel or vegetation as they sink toward the bottom. The average incubation time is about 2 days.

The young fish feed on plankton, then aquatic insects, and later on small fishes. White Bass grow very rapidly under favorable conditions.

Local Names: Silver Bass, Striped Bass

76. Yellow Bass
Morone mississippiensis

Field Marks:
· Body is deep and laterally compressed, deeper than in either the White Bass or White Perch.
· Scales are medium-sized.
· Lateral line is complete, distinct.
· Mouth is medium-sized with the upper jaw ending just ahead of the midpoint of the eye.
· Dorsal fins are connected by a small membrane.

Adult Size: 8–12 in. (20.32–30.48 cm) average

Color: Upper back is olive or brown. The sides are silvery-yellow, and the belly is white or cream. There are 5–7 dark horizontal stripes along the sides; these may be broken into smaller segments below the lateral line. Fins are pale yellow to pale tan.

Habitats: Identical to those of the White Bass.

Natural History: The natural history of the Yellow Bass is very similar to that of the White Bass. Spawning generally occurs in April and May in tributary streams over gravel bottoms. The very small adhesive eggs are broadcast over the spawning areas and sink to the bottom, where they hatch in 4 to 6 days at a temperature of 70°F (21.11°C). The diet consists of plankton at first, followed later by insects, and finally almost entirely of small fishes.

■ SUNFISHES

Order Perciformes
Family Centrarchidae

Field Marks:
· Bodies of members of this strictly North American family, with a few exceptions, are deep and strongly compressed laterally.
· Scales of all species are medium-sized to large. In most species, the cheeks and gill covers are scaled.
· Lateral line is complete and well developed, with the exception of a few species.
· Head is usually small with medium to large eyes.
· Mouths, in most of the small species, are small, while those of the Largemouth and Smallmouth Bass are quite large. Teeth are small and are found on the jaws as well as inside the mouth.
· Seven fins: paired soft-rayed pectorals set low on the sides directly behind the rear of the gill covers; paired ventrals originating about ⅓ of the total body length back from the tip of the lower jaw, often under the pectorals, each ventral fin has a single spine in its leading edge; a single long-based anal originating at or slightly behind the midpoint of the belly, consisting of a spiny forward section and a soft-rayed rear section; a small to medium-sized caudal fin that may be rounded, squared, or slightly indented along its trailing edge; and a single long dorsal fin consisting of a spiny forward section and a soft-rayed rear section.

Habitats: The sunfishes are found in a broad range of freshwater habitats, including lakes, ponds, rivers, and streams. The smaller species prefer warm shallow water near vegetation, rocks, sunken logs, or stumps.

Natural History: The sunfishes are spring and early summer spawners. Most species build well-defined nests, often in densely populated colonies. Males select the nesting site, which may be used year after year. The male is also responsible for guarding the nest, the eggs, and the young.

The smaller sunfishes feed mainly on aquatic insects and crustaceans, but will take nearly any available food. The larger members of the family (the basses) eat fishes, insects, and the larger crustaceans (especially crayfish), reptiles, and small mammals.

Collecting: The sunfishes can be collected in a variety of ways. Minnow seines, minnow traps, and long-handled dip nets will all work under certain conditions. Probably the most enjoyable way to collect sunfishes is with a barbless hook and small live bait on a light fishing rod. Casting from shore or simply dropping the line over the edge of a dock or boat are both productive methods of angling for sunfishes. Small artificial flies tied on barbless hooks will also work very well, especially for the smaller species.

Handling: The sunfishes are excellent aquarium fishes, especially for the beginner. They are hardy, colorful, and abundant in most regions of North America. They are generally undemanding in captivity, requiring only the usual clean, well-aerated water. They will thrive in tanks that are densely planted or those with no plants at all. They tolerate a temperature range from 65°F to near 80°F (15.56°–26.67°C), with the ideal being around 70°F (21.11°C).

Members of this family can be offered many foods that you can easily gather or buy. They will readily take small earthworms, mosquito larvae, small grasshoppers, hellgrammites, and other aquatic insect larvae. Live or dried prepared foods purchased from an aquarium dealer will provide a convenient year-round food supply.

77. Mud Sunfish

Acantharchus pomotis

Field Marks:

· Body is robust and moderately deep.
· Scales are medium-sized. The cheeks and lower parts of the gill covers are scaled.
· Earflaps on the top rear edge of the gill covers are small.
· Mouth is large.
· Anal fin has 5 spines.
· Caudal fin is broad, rounded on its trailing edge.
· First dorsal fin has 12 spines; the second dorsal fin has 1 spine.

Adult Size: 4–6 in. (10.16–15.24 cm) average

Color: The back and sides are light steel gray or greenish-gray. The lower sides and belly are light gray or white. There are 5–8 vague dark horizontal broken lines running from the eyes to the rear edge of the caudal peduncle. Fins are pale gray or very pale olive and are unmarked.

Habitats: The Mud Sunfish is usually found in weedy areas of shallow lowland streams, ponds, and swamps.

Natural History: The Mud Sunfish is shy and secretive. Its natural history is typical of the family, with the male being responsible for nest building and protection of the young.

78. Rock Bass

Ambloplites rupestris

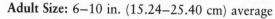

Field Marks:
- Body is moderately deep, moderately laterally compressed.
- Scales are medium-sized. The cheeks and gill covers are scaled.
- Lateral line is complete.
- Anal fin has 5 spines.
- First dorsal fin has 11 spines.
- Caudal fin is broad, indented on its trailing edge.

Adult Size: 6–10 in. (15.24–25.40 cm) average

Color: The upper back is dark brown or dark olive with darker saddle-shaped markings. The sides and belly are dark golden or ochre. There are 8–10 dark horizontal rows along the sides; these are less pronounced in older specimens. The eyes are bright red or orange. Fins are ochre with indistinct brown mottlings or rows of small spots.

Habitats: Rock Bass are usually found near the shorelines of clear, cool lakes, ponds, and large slow-moving streams or rivers. They are often found near weed beds.

Natural History: As water temperatures rise in the spring, male Rock Bass move into shallow water near shore to prepare their nests. Large numbers of fish may gather in one area, and competition for preferred sites may be quite intense. By fanning his tail over the bottom, a male Rock Bass will excavate a nest as large as 2 ft. in diameter and several inches deep at its center. Spawning begins when the water temperature has reached a minimum of 60°F (15.56°C). The ritual begins with the male displaying himself, with fins erect, over his finished nest to attract a female. Once on the nest, the female assumes a head-to-tail position with her mate and slowly rolls over onto one side while the male remains upright. The pair rocks gently back and forth as the eggs are released and fertilized. Some 5000 eggs are released, a few at a time, over a period of about 1 hour. At the end of a spawning the male may drive the female from his nest, and, if she is not already spent, she may move to the nest of another male to finish her spawning while her first mate may spawn with additional females. Spent females leave the nest colony and return to somewhat deeper water. The males remain on the nest to protect the eggs and the fry. One or two weeks after their birth, the male deserts the fry and returns to deeper water.

Rock Bass feed mainly on immature aquatic insects and, to a lesser degree, crayfish and small fishes (which may include their own offspring).

Local Names: Redeye Bass, Redeye, Goggle Eye, Northern Rock Bass

79. Flier

Centrarchus macropterus

Field Marks:
- Body is extremely deep, disc-shaped.
- Scales are medium-sized. The cheeks and gill covers are scaled.
- Anal fin has eight spines and originates just behind the origin of the ventral fins.
- First dorsal fin has 12 spines.
- Caudal fin is broad, slightly indented along its trailing edge, with rounded lobes.

Adult Size: 4–7 in. (10.16–17.78 cm) average

Color: The upper back is dark green or olive. The sides are light olive or light brown shading to pale tan on the lower sides and belly. Each scale is marked with a dark spot, giving the body the appearance of being marked with horizontal rows of dark spots. The dorsal, caudal, and anal fins are pale olive with patches of orange. There is a dark spot at the base of the soft dorsal fin. The ventral and pectoral fins are pale tan.

Habitats: The Flier is found in the acidic water of lowland streams, swamps, and bayous, usually near dense vegetation.

Natural History: The life story of the Flier is typical of its family. Spawning takes place in May and June. The male digs a shallow dishlike nest in shallow water near shore, usually in dense colonies.

The Flier's principal food is insect larvae, especially mosquito larvae. It also feeds on other small aquatic invertebrates, mainly mollusks and crustaceans.

Local Names: Round Sunfish

80. Everglades Pygmy Sunfish
Elassoma evergladei

Field Marks:
· Body is stout, perchlike, oval in cross section.
· Body scales are medium-sized, the scales on the cheeks and gill covers are small.
· Lateral line is absent.
· Ventral fins are long, pointed.
· Anal fin has 3 spines.
· First dorsal fin has 4 spines; the second dorsal fin has 1 spine.
· Caudal fin is distinctly rounded.

Adult Size: 1.5 in. (3.75 cm) maximum

Color: The upper back is dark brown. The sides are buff or tan. The sides are marked by a series of dark brown, irregularly shaped, vertical bars. A few random scales are bright blue or aqua. The fins are tan. The pectoral and ventral fins are unmarked. The soft dorsal, anal, and caudal fins are marked with several rows of dark brown spots.

Habitats: The Everglades Pygmy Sunfish is found in swamps, bayous, and ponds, usually near or in dense vegetation.

Natural History: The Everglades Pygmy Sunfish spawns in late winter and early spring. The male constructs a rudimentary nest of small pieces of plant matter. With fin and color displays the male attracts a female to the finished nest. The eggs hatch in an average of 3 days at 75°F (24.08°C). During the incubation period the male guards and aerates the eggs and remains with the fry until they are old enough to fend for themselves.

This diminutive species feeds mainly on small aquatic insects, small mollusks, and crustaceans.

81. Banded Pygmy Sunfish

Elassoma zonatum

Field Marks:
· Body is stout, only slightly laterally compressed.
· Body scales are medium-sized; the scales on the cheeks and gill covers are small.
· Lateral line is absent.
· Ventral fins are pointed.
· There are 3 short spines in the anal fin.
· Dorsal fin has 5 spines.
· Caudal fin is fan-shaped, rounded on its trailing edge.

Adult Size: 1 in. (2.54 cm) average

Color: The back and top of the head are olive-green. The sides are pale olive or light green and are marked with about 10 darker olive vertical bars. The belly is white or cream. Fins are pale ochre or pale tan. The soft dorsal fin and the caudal fin are marked with several broken rows of dark olive spots.

Habitats: The Banded Pygmy Sunfish spends its time lying hidden under dense mats of vegetation in southern swamps, bogs, and bayous.

Natural History: The Banded Pygmy Sunfish reaches sexual maturity during its first year of life. Spawning takes place in March, when the water temperature has warmed to about 65°F (18.48°C).

The peak spawning season continues through early May. There is no evidence that this species constructs any kind of nest. The eggs are released and fertilized over aquatic vegetation about 1 ft. (0.3 m) above the bottom. The eggs adhere to the vegetation, where they remain during their incubation. The average female produces about 1000 eggs.

The Banded Pygmy Sunfish feeds mainly on aquatic insects but is a sight feeder and will strike anything of appropriate size that moves near it.

82. Blackbanded Sunfish

Enneacanthus chaetodon

Field Marks:
· Body is very deep, almost round.
· Body scales are medium-sized; the scales on the cheeks and gill covers are small.
· There is a very small earflap on the upper rear edge of each gill cover.
· There are 3 spines in the anal fin.
· There are 10 spines in the dorsal fin.
· Rear edge of the caudal fin is rounded.

Adult Size: 2–3 in. (5.08–7.62 cm) average

Color: The ground color of the body and fins is straw or golden. The sides and head are marked with wide irregular black bands. The belly is pale tan, golden, or white. The first three dorsal fin spines and their membranes are velvety black. The other fins, except the pectorals, are marked with rows of small black spots.

Habitats: The Blackbanded Sunfish lives in shallow, tea-colored, quiet water in streams, the back waters of rivers, lakes, and ponds. It is generally found near dense vegetation. It is absent from many waters within its range that appear suitable to its lifestyle.

Natural History: The prespawning and mating activity of this species is typical of the family. The males dig small nests in shallow water near shore, usually on a sand bottom. The incubation period averages 2–3 days.

The Blackbanded Sunfish does most of its feeding after dark, consuming mostly immature aquatic insects and plant matter.

83. Bluespotted Sunfish
Enneacanthus gloriosus

Field Marks:
· Scales on the body are medium-sized; the scales on the cheeks and gill covers are small.
· Lateral line is incomplete.
· Earflap is small.
· There are 3 spines in the anal fin.
· There are 8 spines in the dorsal fin.
· Caudal fin is broad, round on its trailing edge.

Adult Size: 3–4 in. (7.62–10.16 cm) average

Color: The back ranges from very dark olive to almost black. The sides are lighter olive or slate gray. The belly is pale gray and the underbelly is dusky. The sides are marked with irregular horizontal rows of light blue or aqua. There is a small black spot on the ear flap. The fins are light gray to dusky. The dorsal, caudal, and anal fins are marked with rows of light blue or aqua spots radiating outward from their bases.

Habitats: The Bluespotted Sunfish is most often found in shallow weedy lowland streams and ponds with a low pH.

Natural History: The male Bluespotted Sunfish digs a small dish-like nest, 6–12 in. (15.24–30.48 cm) in diameter, usually in very shallow weedy water near shore. The nests may be dug in mud, silt, sand, or gravel. The fertilized eggs are adhesive and attach to the vegetation or substrate of the nest.

Bluespotted Sunfish feed on immature aquatic insects and small crustaceans.

84. Banded Sunfish
Enneacanthus obesus

Field Marks:
- Body scales are medium-sized; the scales on the cheeks and gill covers are small.
- Eye is large.
- Earflap is medium-sized.
- Anal fin, which originates just behind the ventral fins, has 3 short spines.
- Dorsal fin has 9 spines.
- Caudal fin is broad, nearly straight on its trailing edge.

Adult Size: 3–3.5 in. (7.62–8.75 cm) maximum

Color: The upper back and top of the head are dark slate gray or steel blue. The sides are tan and marked with 5–8 dark, well-defined, vertical bars. The chin and breast are white or tan. The belly and ventral surfaces are pale purple. Fins are light brown. The ventral and pectoral fins are unmarked. The caudal, anal, and dorsal fins are spotted with darker brown. The spiny dorsal fin is marked with light blue spots.

Habitats: The Banded Sunfish is most often found in warm, weedy, lowland ponds and streams, often in association with the Redfin Pickerel.

Natural History: What little information is available on this species suggests that its biology is very similar to that of the Blue-spotted Sunfish (p. 143).

Banded Sunfish feed on small aquatic insects and crustaceans.

Local Names: Barred Sunfish

85. Redbreast Sunfish

Lepomis auritus

Field Marks:
· Body is moderately deep.
· Scales on the body are medium-sized; those on the cheeks and gill covers are small.
· Earflap is large.
· Anal fin has 3 spines.
· Dorsal fin has 10 or 11 spines.
· Caudal fin is indented on its trailing edge and has rounded lobes.

Adult Size: 5–8 in. (12.7–20.32 cm) average

Color: The back is dark or medium brown above. The sides are tan or pale yellow. The chin, breast, and belly are yellow or orange (red in breeding males). The sides are marked with irregular dark brown or olive vertical bars. There are several aqua lines around the eye and the snout. The earflap is intense dark blue or black. Fins are tan (pectorals, ventrals, and anal) or ochre (dorsal and caudal). The soft dorsal fin is marked with a row of dark brown spots at its base. The caudal fin has a band of dark brown slightly ahead of its trailing edge.

Habitats: The Redbreast Sunfish prefers warm shallow water near vegetation over a rock, sand, or gravel bottom in lakes, ponds, and slow-moving rivers or streams. It is often found in association with the Rock Bass.

Natural History: The Redbreast Sunfish spawns somewhat later than most members of its family. The males, which are very brightly colored in reds and oranges during the spawning season, prepare shallow nests that may be as large as 2–3 ft. (0.61–0.91 m) in diameter. The nests are usually built in colonies of 5–10, in water ranging from 6 to 18 in. (15.24 to 45.72 cm) deep. The Redbreast Sunfish may occasionally use the abandoned nests of other members of its family, including those of the Largemouth Bass. The male attracts a mate to the nest by fin and color displays. The fertilized eggs are adhesive.

The diet consists mainly of immature aquatic insects but may include other aquatic invertebrates and small fishes.

Local Names: Longear Sunfish, Longears, Yellowbelly Sunfish

86. Green Sunfish
Lepomis cyanellus

Field Marks:
- Body is moderately deep.
- Body scales are medium-sized; the scales on the cheeks are small, on the gill covers medium-sized.
- Earflap is large.
- Anal fin has 3 short spines.
- Dorsal fin has 9–11 (usually 10) spines.
- Caudal fin is broad, indented on its trailing edge with rounded lobes.

Adult Size: 5–8 in. (12.7–20.32 cm) average

Color: The back is olive-green to olive-brown. The sides are light green or tan and are marked with a series of dark brown or green vertical bars. The breast is usually dull orange or yellow. The earflap is black with a red or orange margin. There are a few blue or aqua spots around the eyes. The fins are olive to brown and are usually unmarked.

Habitats: Green Sunfish are found in a broad range of habitats, from clear, cool lakes and ponds to small turbid streams and drainage ditches. Often found near submerged brush piles, sunken logs, and dense vegetation.

Natural History: The Green Sunfish is typical of its family in most respects. The males are responsible for constructing the nests, which they do in shallow water near shore. They are much less likely to nest in colonies than are most of the other small sunfishes. The nests are usually placed on rocky bottoms. Often more than one female will spawn in each nest. The fertilized eggs are adhesive and require 3–5 days to hatch. The male drives his mate from the nest after spawning, and he remains on the nest for about a week after spawning to protect the eggs and the young until they are free-swimming. Redfin Shiners are known to spawn in and near the nests of Green Sunfish. The Bluegill and Green Sunfish often interbreed, resulting in hybrid offspring.

The young fish grow very rapidly during their first year, feeding (as do the adults) on aquatic and terrestrial insects and, to a lesser extent, on mollusks and small fishes.

Local Names: Green Perch, Sand Bass, Black Perch

87. Pumpkinseed
Lepomis gibbosus

Field Marks:
- Body is very deep, strongly compressed laterally, almost disc-shaped.
- Scales on the body are medium-sized, on the cheeks and gill covers they are small.
- Earflap is medium-sized.
- Anal fin has 3 spines.
- Dorsal fin has 10–11 (usually 10) spines.
- Trailing edge of the caudal fin is moderately indented with rounded lobes.

Adult Size: 4–9 in. (10.16–22.86 cm) average

Color: The back is medium to dark brown. The sides range from tan to pale yellow. The chin and breast are yellow or orange. The sides are marked with a series of dark, indistinct vertical bars. There are 5–7 light blue or aqua lines that radiate back from the eye and continue across the gill covers and onto the body. The earflap is black or dark blue with a red rear edge. Fins range from ochre to tan or yellow. The caudal and anal fins may be marked with rows of small dark brown spots. The rear edges of the dorsal, caudal, and anal fins have a thin border of aqua.

Habitats: Found in a broad range of habitats, in warm weather the Pumpkinseed prefers shallow water near shore, usually near vegetation, or over rock, sand, or gravel bottom free of vegetation. It is commonly found in lakes, ponds, rivers, and the slower parts of streams or brooks. In winter the Pumpkinseed moves into deep water, where it forms dense schools known as winter aggregations.

Natural History: When the water has reached a temperature of about 68°F (20.16°C), male Pumpkinseeds move into the spawning areas and begin work on the nests, which are often used for many consecutive spawning seasons and may need only a cleaning to prepare them for spawning. Usually arranged in dense colonies, the nests are placed in shallow water near shore with an average depth of 8–16 in. (20.32–40.64 cm). The completed nests have an average diameter of 5–15 in. (12.70–38.10 cm). Mating pairs swim over the completed nests while they nip and nuzzle one another; this behavior continues throughout the spawning act. Shortly before the release of the eggs, the female turns onto her side while the male remains upright. Both eggs and sperm are released in small quantities at irregular intervals. The male does not drive his mates from the nest, although he may take another mate if one of his should move on to another male. The amber colored eggs are adhesive after they are fertilized. The male remains on the nest after spawning, fanning the eggs to ensure that they have an adequate supply of oxygen and are kept free of silt and other debris. At a water temperature of 80°F (26.67°C) the eggs require 3 days for complete incubation; the incubation period is increased at lower temperatures. The male stays on the nest until the fry can care for themselves, usually 7–12 days after spawning. After the first brood has left the nest, the male may prepare the nest for another spawning. The spawning season may continue through July.

Both young and adult Pumpkinseeds feed mainly on immature aquatic insects. They also consume aquatic worms and mollusks and an occasional snail or small fish. They are sight feeders and actively search for food throughout the day and early evening.

Local Names: Yellow Sunfish, Sunny, Kibbie, Sun Bass

88. Warmouth

Lepomis gulosus

Field Marks:
- Body is moderately deep.
- Body scales are medium-sized, the scales on the cheeks and gill covers are smaller.
- Mouth is large.
- Anal fin has 3 spines.
- Dorsal fin has 10 spines (rarely 9 or 11).
- Caudal fin is slightly forked with rounded lobes.

Adult Size: 4–10 in. (10.16–25.4 cm) average

Color: The back and top of the head are dark brown or olive. The sides are ochre or golden and are marked with dark, irregular, broken vertical bands. The belly and chin are golden or ochre. The cheeks and gill covers are marked with several dark brown stripes radiating back from the snout and eye. Fins are ochre or brown. The dorsal and caudal fins are marked with rows of dark brown spots.

Habitats: The Warmouth prefers shallow water near dense vegetation, especially in the backwaters of large, slow-moving rivers. Often found in small pockets of open water surrounded by vegetation near submerged stumps, logs, or brush.

Natural History: The spawning season for this species begins in May and may continue into August, with its peak being the month of June. Males select nest sites over soft mud or silt bottoms, either in deep or shallow water. The nests are often situated at the base of sunken logs or stumps. The spawning act is typical of the sunfish family: the male prepares the nest, courts and mates with one or more females, drives his mates from the nest after spawning, and jealously guards the nest. Intruders into the male's territory are greeted by his rushing at them with his gill covers flared, his mouth open, his eyes turning bright red, and the body color turning from brown to bright yellow. The incubation period averages about 3 days, and the fry leave the nest about 5 days after hatching.

Young Warmouth feed mainly on aquatic crustaceans, such as water fleas. The adults feed on crayfish, immature aquatic insects, and small fishes.

Local Names: Mud Bass, Stumpknocker, Goggle-eye

89. Orangespotted Sunfish
Lepomis humilis

Field Marks:
- Body is moderately deep.
- Body scales are medium-sized. The scales on the cheeks and gill covers are small.
- Earflap is large.
- Anal fin has 3 short spines.
- Dorsal fin has 10 (rarely 9 or 11) spines.
- Caudal fin is slightly forked with rounded lobes.

Adult Size: 3–4 in. (7.62–10.16 cm) average

Color: In nonbreeding males and females, the back is dark olive or brown. The sides are light green with blue reflective surfaces. The lower sides are marked with reddish-brown spots. The earflap is black with a white margin. The fins are light brown or tan and are unmarked.

Breeding males have several brown saddle-shaped marks along their upper backs. The sides are tan and are marked with bright orange spots. There are several orange streaks on the cheeks and gill covers. The belly and breast are orange or red. The chin is yellow or orange.

The ventral, anal, and dorsal fins are orange. The pectoral and caudal fins are very pale tan. The caudal may be marked with a few rows of dark brown spots along its base.

Habitats: The Orangespotted Sunfish prefers shallow, often silted, water near weed beds, large rocks, sunken logs, or submerged brush piles.

Natural History: The Orangespotted Sunfish begins spawning earlier than do most of the sunfishes, usually in late April or early

May. The spawning season may continue into August. The nests are built by the males in shallow water, usually near shore. The male remains on the nest until the eggs hatch, normally about 5 days. Redfin Shiners and Red Shiners are known to use the nests of Orangespotted Sunfish for their own spawning.

The diet of this small species is made up of immature aquatic insects, crustaceans, and occasional small fishes.

90. Bluegill

Lepomis macrochirus

Field Marks:
- Body is very deep.
- Scales on the body are medium-sized. The scales on the cheeks and gill covers are small.
- Earflap is medium-sized.
- Pectoral fins are long, pointed.
- Anal fin has 3 spines.
- Dorsal fin has 10 (rarely 11) spines.
- Caudal fin is indented along its trailing edge with slightly rounded triangular lobes.

Adult Size: 4–10 in. (10.16–25.4 cm) average

Color: The back is dark olive to dark brown. The sides are dull purple or dark orange. The sides are marked with a series of dark vertical bands. The breast and belly are brown or orange. There are several wavy blue or aqua lines on the chin, the lower parts of the gill covers, and the lower parts of the cheeks. In breeding males, the breast and belly become bright coppery orange. Fins range from light olive to tan or light gray-green. The fins are unmarked except for a dark spot at the base of the soft dorsal and a few small spots on the caudal.

Habitats: The Bluegill is most common in warm-water habitats in shallow weedy ponds, lakes, and slow-moving streams. It shows a preference for weedy water near shore.

Natural History: The biology of the Bluegill is very similar to that of the Pumpkinseed (p. 147). Bluegills move out of their winter aggregations when the water temperature has reached 50°F (10°C). The males move onto the spawning areas, in shallow water near shore, where they claim and defend nesting territories. Nests may be used for many consecutive seasons. They are usually crowded into dense colonies in 2–3 ft. (0.61–0.91 m) of water. The nests average 18–24 in. (45.72–60.96 cm) in diameter, and one nest may be used by several different pairs of fish during the spawning season (from late June through August). It is common for several females to mate with one male over the same nest, resulting in most nests containing a large number of fertilized eggs. The adhesive eggs are guarded and aerated by the male during their 3–5-day incubation period. After the fry leave the nest, the male may prepare the nest for another spawning. The Bluegill may hybridize with several other species of sunfish, including the Redbreast, Pumpkinseed, and Green. Where this happens, identification may be very difficult.

Bluegills feed mainly on aquatic mollusks, worms, and immature and adult aquatic and terrestrial insects. They also occasionally consume plant matter.

Local Names: Blue Sunfish, Bream, Roach, Pond Perch

91. Longear Sunfish

Lepomis megalotis

Field Marks:
- Body is very deep.
- Scales on the body are medium-sized. Scales on the gill covers and the cheeks are small.
- Earflap is large.
- There are 3 spines in the anal fin.
- Dorsal fin has 10 (rarely 9 or 11) spines.
- Caudal fin is broad, indented along its trailing edge with rounded lobes.

Adult Size: 3–6 in. (7.62–15.24 cm) average

Color: Nonbreeding males and females have olive or brown backs and blue-green sides marked with yellow and pale green spots and irregular olive or brown vertical bands. The belly is yellow or orange. The head is olive or light orange with aqua or pale blue wavy lines. The earflap is black with a narrow white border. In breeding males the vertical bars on the sides are dark blue on a dark orange ground that brightens to intense orange or yellow on the belly. The ground color of the head also becomes intense orange or yellow.

Fins of nonbreeding males and females are pale tan and are usually unmarked. In breeding males, the ventral fins are deep indigo blue. The other fins are orange. The dorsal and caudal fins may be marked with rows of blue or aqua spots.

Habitats: Nearly always found in shallow water, the Longear Sunfish prefers habitats with dense vegetation in small lakes, ponds, and slow-moving rivers and streams.

Natural History: The natural history of the Longear Sunfish is very similar to that of the Green Sunfish. The Longear spawns from May through late July or early August in nests whose edges may touch, arranged in dense colonies near shore. The male digs the nest and defends his territory during the prespawning period. Females approach the nesting colony when the nests are completed. Vertical bars on the females' sides become very intense as they swim over the nest. The males greet their mates by swimming around and above them tilted to one side to display their bright orange bellies. The mating pairs swim in circles over the nests, stopping for short intervals to release small amounts of eggs and sperm. At any time during spawning, the male may chase his mate from the nest; she may return to spawn with the same male or move on to another mate.

Longear Sunfish feed on small aquatic invertebrates, insects, and, occasionally, small fishes.

Local Names: Northern Longear, Great Lakes Longear, Pumpkin-seed, Creek Perch

92. Redear Sunfish

Lepomis microlophus

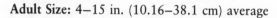

Field Marks:
- Body is deep, disc-shaped.
- Scales on the body are medium-sized. The scales on the cheeks and gill covers are small.
- Lateral line is complete.
- Earflap is small.
- Pectoral fin is long, pointed.
- Anal fin has 3 spines.
- Dorsal fin has 10 (rarely 9 or 11) spines.
- Caudal fin is broad, slightly indented on its trailing edge with rounded lobes.

Adult Size: 4–15 in. (10.16–38.1 cm) average

Color: The upper back is slate gray or dusky continuing down the sides in 5–10 indistinct vertical bars. The ground color of the sides is pale slate gray or slate blue. The breast and chin are very pale to intense orange or yellow. The earflap is black with a red border. Fin membranes are dark brown contrasting with tan rays, except on the spiny dorsal fin where the reverse is true. There is a small dark crescent at the base of the pectoral fins.

Habitats: Generally found in larger bodies of water than most of the other small sunfishes, the Redear Sunfish congregates around bottom structures, like logs, stumps, large rocks, and submerged brush piles. It is less fond of dense vegetation than are many of the other small sunfishes.

Natural History: The Redear Sunfish differs from the typical member of this family in that it grows larger and is less prolific than most of the other small sunfishes.

The spawning behavior of the Redear Sunfish is identical to that of the Bluegill (p. 151) and Longear (p. 152) Sunfishes. Spawning takes place in May and June and occasionally in August.

The Redear Sunfish uses the flat-topped teeth in its throat to crunch the shells of snails, which are its favorite food. If snails are readily available, you should feed them to captive Redear Sunfish. They can also be trained to accept other live foods and dried foods.

Local Names: Shell Cracker, Stumpknocker, Yellow Bream

93. Spotted Sunfish

Lepomis punctatus

Field Marks:
- Body is very deep.
- Scales on the body are medium-sized; on the cheeks and gill covers, they are small.
- Earflap is small.
- Anal fin has 3 spines.
- Dorsal fin has 10 (rarely 9 or 11) spines.
- Broad caudal fin is indented on its trailing edge and has rounded lobes.

Adult Size: 3–6 in. (7.62–15.24 cm) average

Color: The back is dark olive or dark brown. The sides are light green or light olive. The lower parts of the sides and the belly are orange or yellow. Each body scale is marked with a small dusky or black spot. Fins are dusky olive. The upper portion of the soft dorsal fin is dull orange.

Habitats: Usually found near weeds, logs, large rocks, or other sheltering structures, the Spotted Sunfish occurs in slow-moving lowland streams and warm, shallow ponds.

Natural History: The spawning season of the Spotted Sunfish begins in May or June and continues well into July and, perhaps, August. The nests are usually not built in colonies; although, on occasion, several nests will be placed very close together. The courting and spawning behavior is typical of the sunfish family (see accounts of other species, i.e., Pumpkinseed, pp. 147–148).

The eggs hatch in about 4 days at a water temperature of 75°F (24.08°C). The fry leave the nest about 10 days after hatching.

Spotted Sunfish feed on immature and adult aquatic insects, mollusks, and crustaceans.

94. Bantam Sunfish

Lepomis symmetricus

Field Marks:
- Body is deep and disc-shaped.
- Body scales are medium-sized; on the cheeks and gill covers, they are small.
- Earflap is medium-sized.
- Anal fin has 3 spines.
- Dorsal fin has 10 (rarely 9 or 11) spines.
- Caudal fin is broad, slightly indented on its trailing edge with rounded lobes.

Adult Size: 2–3 in. (5.08–7.62 cm) average

Color: The dark brown of the upper back extends down over the sides in a series of vertical bars (usually 5–7). Between the bars, the sides are light tan or light grayish-blue. The belly is orange. The earflap is black with a partial thin white edge. The cheeks and gill covers are light blue or aqua. Fin membranes are brown, contrasting with tan rays, except on the spiny dorsal, where the reverse is true. There is a distinct black spot on the rear of the soft dorsal fin.

Habitats: The Bantam Sunfish is found in warm, weedy lowland ponds and sluggish streams.

Natural History: Bantam Sunfish spawn in late spring and early summer. Their biology, including spawning habits, is typical of the sunfish family.

Its primary food is immature aquatic insects and crustaceans.

95. Smallmouth Bass

Micropterus dolomieui

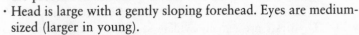

Field Marks:

- Body is elongated, robust, slightly compressed laterally.
- Scales on the body are medium-sized; on the cheeks and gill covers, they are slightly smaller.
- A few rows of very small scales are present on the bases of the soft dorsal and anal fins.
- Head is large with a gently sloping forehead. Eyes are medium-sized (larger in young).
- Mouth is large. The upper jaw extends to the midpoint of the eye. The lower jaw projects beyond the upper.
- Pectoral fins are rounded.
- Anal fin has 3 spines.
- Spiny dorsal fin has 10 spines; the soft dorsal fin has 1 spine.
- The caudal fin is broad, slightly forked, with rounded lobes.

Adult Size: 6–18 in. (15.24–45.72 cm) average

Color: The back is usually dark brown or dark olive, sometimes nearly black. The sides are greenish-yellow to bronze. The chin, breast, and belly may be white, cream, or pale yellow. The sides are marked with 9–12 dark brown or olive indistinct vertical bars; these bars are irregular and may appear only as a mottled pattern. The back is usually marked with 9–12 diamond-shaped spots. The cheeks and gill covers are marked with 4–5 dark stripes radiating out from the eyes. The eyes are bright red or orange.

The pectoral and anal fins are tan to light brown and are unmarked. The dorsal and caudal fins are brownish-yellow or light olive. The base of the soft dorsal fin may be marked with small dark spots. The ventral fins are usually yellow.

All the colors are lighter in juvenile specimens. The caudal fin is coral color with a dark crescent. The other fins are pale tan or pale coral. The soft dorsal fin has a dark crescent near its trailing edge.

Habitats: A fish of cool, clear lakes, ponds, and rivers, the Smallmouth Bass is usually found over gravel bottoms near large rocks, logs, or stumps.

Natural History: As the spawning season approaches, male Smallmouth Bass move out of their winter aggregations and into shallow water along rocky points, bars, gravel shorelines, or ledges near deep water. Spawning takes place in spring and early summer. On hard bottoms the nests may be little more than an area from which the silt and debris have been fanned away; on softer bottoms, like sand, they are dishlike depressions that average 1–4 ft. (0.30–1.22 m) in diameter. The fish usually return to the same nesting areas each year, reusing the same nest for many years. The nests are usually widely separated, not clustered together as are the smaller sunfishes'.

When the nest is finished, the male hovers over it and displays his intensified spawning colors. Once a mate has been attracted onto the nest, the pair begins swimming quickly around the nest rubbing against one another, nipping, and showing fin and color displays. When they are ready to spawn they come to rest in the center of the nest and the female turns onto her side while the male remains upright. The eggs and sperm are released in a series of short bursts over a period of time up to 2 hours long. Usually the male will drive his first mate from the nest before she is spent. Both sexes normally spawn with more than one mate.

The male remains on the nest after spawning and guards and fans the eggs, which usually lie in a mass at the nest's center. The eggs hatch in an average of 4–10 days. About 6 days after hatching, the fry begin to leave the nest; but the father returns the strays to the nest by carrying them in his mouth and guards and herds them for several more days. After the fry leave the nest the male may clean it in preparation for additional spawning.

Feeding first on plankton and later on aquatic insects, the fry grow rapidly, often reaching a length of 2–4 in. (5.08–10.16 cm) during their first summer. Adult Smallmouth Bass feed mainly on insects, crayfish, and fish. They may also take frogs, mice, or small snakes.

Local Names: Northern Smallmouth Bass, Black Bass, Bronzeback, Smallmouth, Redeye Bass, Brown Bass

96. Largemouth Bass
Micropterus salmoides

Field Marks:
· Body is elongated, robust, deeper than the Smallmouth Bass; slightly compressed laterally.
· Scales on the body are medium-sized; smaller on the cheeks and gill covers.
· Head is very large; eyes medium-sized in adults, large in young.
· Mouth is very large; the upper jaw extends beyond the rear edge of the eye. The lower jaw projects beyond the upper.
· Anal fin has 3 spines.
· Spiny dorsal fin has 10 spines, the last spine appears to be part of the soft dorsal fin. There is a deep notch between the spiny and soft dorsal fins, although they are not separated.
· Caudal fin is broad, slightly indented on its trailing edge with rounded lobes.

Adult Size: 8–18 in. (20.32–45.72 cm) average

Color: The back is dark olive to dusky or black. The sides are yellowish-green or light olive. The belly is white or creamy. A broad dark (usually black) lateral band begins at the eye and ends, in young specimens, on the caudal fin, in adults, at the rear edge of the caudal peduncle. This band fades with age and may be solid or broken. The back is usually marked with dark saddles. There may be narrow dark stripes on the cheeks and gill covers. Dorsal and caudal fins are pale tan or pale olive. The pectoral and ventral fins are very pale tan or whitish. The anal fin is white or tan. In young fish there are dark spots on the caudal and soft dorsal fins.

Habitats: Largemouth Bass are most often found in shallow, weedy water, seldom far from dense vegetation or other cover. They are common in lakes, ponds, and sluggish rivers and streams and show a distinct preference for warm water.

Natural History: Largemouth Bass begin spawning in late May or early June and may continue into August. The peak of the spawning season is usually from early to mid-June. Males become very aggressive as they claim territories and work on building nests. The nests are usually placed in the shallow weedy waters of small coves and bays. Completed nests range from 2–3 ft. (0.61–0.91 m) in diameter and may be as much as 8 in. (20.32 cm) deep. Many nests may be found in the same general area but will usually be no closer than 30 ft. (9 m) apart. Usually the female is driven from the nest during the spawning before she is spent, and moves on to another mate while the male attracts another female to the nest and mates with her. The male remains on the nest to guard the eggs, which are usually spread over the entire bottom of the nest. During this time Golden Shiners and Lake Chubsuckers may spawn in the Largemouth Bass's nest.

When the fry hatch and have absorbed their yolk sacs, they rise from the bottom of the nest and form a dense school; they remain in this school for as long as a month. The father continues to protect them for at least a few days and occasionally stays with them until the school has dispersed.

The fry begin feeding on plankton and then begin to take increasing amounts of immature aquatic insects. By the time they are about 2 in. (5.08 cm) long, their diet consists mainly of small fishes. Adult Largemouth Bass are sight feeders, devouring anything that comes their way. They have been known to eat frogs, crayfish, snakes, small mammals, and an occasional bird. The bulk of their diet is small fish.

Local Names: Green Bass, Black Bass, Northern Largemouth Bass, Lineside Bass

97. White Crappie
Pomoxis annularis

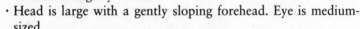

Field Marks:
- Body is moderately deep, more elongated than the Bluegill (p. 151) and deeper than the Smallmouth Bass (p. 157).
- Body scales are medium-sized; the scales on the cheeks and gill covers are slightly smaller.
- Head is large with a gently sloping forehead. Eye is medium-sized.
- Mouth is large with the upper jaw ending at about the rear edge of the eye.
- Pectoral fins are long and rounded.
- Anal fin has 5 spines.
- Dorsal fin has 6 spines.
- Caudal fin is long, slightly indented along its trailing edge, with rounded lobes.

Adult Size: 5–12 in. (12.7–30.48 cm) average

Color: The upper back is dark green, olive, or dark brown. The sides are yellowish-green, shading to cream or white on the belly. There are 5–10 indistinct green vertical bars on the sides. The ground color of the fins is pale cream or light yellow. The anal, caudal, and dorsal fins are marked with several rows of gray, olive, or dusky spots radiating out from their bases.

Habitats: The White Crappie is a schooling fish and often congregates around sunken brush, logs, and stumps in warm, shallow, weedy areas of lakes, ponds, and slow-moving, silted rivers and streams.

Natural History: In typical sunfish family fashion, when the late spring spawning season draws near (usually in late May or early June), the male White Crappie moves onto the nesting grounds and begins clearing an area some 10–12 in. (25.4–30.48 cm) in diameter. The nests are cleared among aquatic vegetation in 10–30 in. (25.4–76.2 cm) of water near shore. The nests may be in colonies or well separated from one another. Usually several plants will be left within the confines of the nest.

Spawning takes place at a water temperature of 60°–75°F (15.56°–24.08°C). The spawning act involves little of the circular swimming motion employed by other sunfishes. The eggs are released and fertilized in as many as 50 short bursts separated by resting periods of as much as 20 minutes. The pair usually changes partners one or more times before the end of the spawning. The colorless eggs are adhesive after fertilization and cling to the plants in the nest or the substrate. The incubation period is usually 4 days. The male tends the eggs, fanning them with his fins to prevent their being suffocated by silt or debris.

White Crappies are quite prolific and may so overpopulate a body of water that only stunted fish remain.

The young feed on plankton, changing their diet as they grow to include an increasing percentage of immature and adult aquatic insects. The adults consume large quantities of small fishes. They feed most actively in the early morning and again in the late afternoon and evening.

Local Names: Silver Bass, Crawpie, White Bass, Papermouth

98. Black Crappie
Pomoxis nigromaculatus

Field Marks:
· Body is deeper than that of the White Crappie, similar to the body of the Pumpkinseed (p. 147).
· Scales on the body and gill covers are medium-sized; on the cheeks they are slightly smaller.
· Head is large. Eyes are medium-sized.
· Mouth is large, with the upper jaw ending at about the midpoint of the eye. Lower jaw projects beyond the upper.
· Pectoral fin is long.
· Anal fin has 6 spines.

· Dorsal fin has 7 spines.
· Caudal fin is long, slightly forked, with rounded lobes.

Adult Size: 4–11 in. (10.16–27.94 cm) average

Color: The upper back is dark olive to dark brown. The sides are pale green or pale tan. The breast and belly are pale tan to cream or light gray. The back and sides are marked with irregular broken rows of gray, dusky, or black spots. Pectoral and ventral fins are very pale tan to light yellow and are unmarked. The dorsal, caudal, and anal fins are pale tan or olive and are strongly marked with rows of gray, brown, olive, or black spots.

Habitats: The Black Crappie is found in large lakes, ponds, and slow-moving streams, where it forms loose schools around plants, sunken logs, or brush piles. It generally prefers cooler, deeper, clearer waters than does the White Crappie.

Natural History: The habits of the Black Crappie are very similar to those of its closest relative, the White Crappie. It usually begins its spawning season in late May and continues it into the middle of July. The male clears a roughly circular nest, which may be no more than a clearing on the bottom or a shallow depression in the substrate. Small plants are usually left within the perimeter of the nest. The nests are usually placed in water 1–2 ft. (0.3–0.61 m) deep under overhanging banks, wherever these are available. Both males and females will normally mate with several partners before they are spent. The male guards the eggs during their 3–5 day incubation period and for several days after they have hatched.

The young fish feed on tiny crustaceans and immature aquatic insects until about their third year. As adults, they feed mainly on small fishes.

Local Names: Speckled Bass, Calico Bass, Shiner, Moonfish, Crawpie, Strawberry Bass, Grass Bass

■ PERCHES

Order Perciformes
Family Percidae

Field Marks:

- Body is usually elongated, but may be stout and robust. It is usually slightly compressed laterally.
- Body scales range from small to medium-sized. In some species, parts of the head are scaled, in others there are no scales on the head.
- Lateral line is usually complete, but may be incomplete or absent.
- Head may be medium-sized or large, with a pointed, rounded, or squared snout. Eyes are medium-sized or large.
- Mouths range in size from very small to large and may be terminal or inferior. There are usually well-developed teeth on the jaws and in the mouth.
- Eight fins: paired pectorals (often broad—fan-shaped, especially in the darters), set just behind the rear edge of the gill covers; paired ventrals that originate just below the pectorals have 1 spine in their leading edge; a single, usually short-based anal with 1 or 2 spines; a caudal that may be squared, rounded, or forked; and 2 widely separated dorsals, the first with 6–15 spines, the second with 1 or 2 spines.

Habitats: Members of the perch family occur in a broad range of habitats, from muddy pools and ditches to clear cool lakes, ponds, rivers, and streams.

Natural History: Members of this family spawn early in the spring. The spawning ritual may be very elaborate, with brightly colored males engaging in intricate mating dances; or it may be very casual, with the mating pairs simply scattering the eggs over vegetation.

The diets of the perches include a wide variety of foods, ranging from tiny plankton to small fishes.

Collecting: Almost any fishing method will work for the perches as long as it is appropriate for the collecting site and the size of the fish. Minnow seines and traps or large dip nets are good tools for catching darters (which are in the perch family); the seine is probably the best under most conditions. Yellow Perch can be taken by any method you choose.

Handling: Some of the darters require rather specific conditions in the aquarium; others require no special care. Duplicating a species' natural habitat as far as possible is the best way to insure the health of your fish.

99. Eastern Sand Darter
Ammocrypta pellucida

Field Marks:
- Body is elongated, slender, and delicate.
- Body scales are small, weakly developed except along the lateral line and may be absent on belly and back. The cheeks and gill covers are not scaled.
- Lateral line is usually complete, but may be incomplete.
- Snout is bluntly rounded. Eyes are large and set high in the head.
- Mouth is very small, terminal.
- Pectoral and ventral fins are long.
- First dorsal fin has 8–11 weak spines.
- Second dorsal fin has no spines.
- Caudal fin is small, almost square.

Adult Size: 2–2.5 in. (5.08–6.25 cm) average

Color: The head and body are light straw-colored or light ochre with medium brown on the upper back and cream on the belly. There is a row of dark olive spots along the upper back and a row of light olive spots along the lateral line. Fins are nearly transparent or very pale tan. The second dorsal and caudal fins are slightly darker along their trailing edges.

Habitats: The Eastern Sand Darter occurs over sand bottoms in streams, rivers, ponds, and lakes with clear, cool water.

Natural History: Spawning occurs in the spring in shallow riffles of small streams. When the males have made the nests ready, ripe females will move into the nesting territories and select mates. Spawning takes place when the female burrows into the bottom, making a small trench where she deposits a few eggs while the male lies immediately beside her and fertilizes the eggs as they are released. This process continues until the female is spent. The male guards the nest area for a short time after spawning.

The Eastern Sand Darter spends most of its time buried in the sand with only its eyes exposed. From this concealed position it springs up to intercept its prey. It feeds mainly on small aquatic crustaceans and immature aquatic insects.

100. Greenside Darter

Etheostoma blennioides

Field Marks:
- Body is elongated and robust.
- Scales on the body, cheeks, and gill covers are small.
- Lateral line is nearly straight.
- Head is large with a squared snout. Eyes are large and placed very high on the sides of the head.
- Mouth is medium-sized, inferior.
- Pectoral fins are large and fanlike.
- Anal fin has 2 spines.
- First dorsal fin has 12 or 13 strong spines.
- Caudal fin is nearly square.

Adult Size: 2.5–3 in. (6.25–7.62 cm) average

Color: The ground color is olive-green or olive-brown, darker above and lighter below. There is a series of dusky saddle-shaped marks along the upper back. Above the lateral line is a row of irregularly shaped dark spots. Below the lateral line are 5–7 dark V-shaped marks. Two dark stripes radiate down and forward from the eye. In breeding males, the markings are intensely bright green. Dorsal fins have a row of olive spots along their bases. The second dorsal and caudal fins are marked with a few rows of long brown spots. The ground color of the dorsals and caudal fins is light olive or green. The pectoral fins are nearly transparent. The ventral and anal fins are very pale tan.

Habitats: The Greenside Darter is found in the tributary streams of large lakes and rivers.

Natural History: Greenside Darters begin spawning when the water temperature has risen above 55°F (13.1°C), usually from April into June. As the season approaches, the males select and defend small territories in shallow water over a gravel bottom. However, it is the females that select the nest sites. The females remain downstream from the males until their eggs become fully ripe, when they move onto the spawning area and locate a rock or log. When a female is satisfied with her choice of nest sites, she turns onto one side. A male will approach her and swim into a mating stance on her back. The pair vibrates as the eggs and sperm are released in short bursts. The fertilized eggs are adhesive and require about 18–20 days for complete incubation.

The bulk of the diet for both young and adult Greenside Darters is immature aquatic insects.

101. Rainbow Darter

Etheostoma caeruleum

Field Marks:
· Body is elongated, robust.
· Scales on the body and gill covers are medium-sized. Cheeks are not scaled.
· Lateral line is complete.
· Head is large with a blunt, rounded snout.
· Mouth is medium-sized, inferior, slightly overhung by snout.
· Pectoral fins are long, fan-shaped.
· Anal fin has 2 spines.
· First dorsal fin has 10 (rarely 9 or 11 spines). Second dorsal fin is higher than the first.
· Caudal fin is long, nearly square.

Adult Size: 2–2.5 in. (5.08–6.25 cm) average

Color: Females and nonbreeding males are green with blue or gray vertical bars on the sides. Breeding males are brick red or red-orange with intense dark blue vertical bars. The head is always green. Fins of females and nonbreeding males are usually pale green or greenish-blue. The fins of breeding males are bright reddish-orange with intense blue markings; the ventral fins in breeding males are solid blue.

Habitats: Rainbow Darters are commonly found in clear cool streams with ample shallow riffles.

Natural History: In typical darter fashion, male Rainbow Darters move onto the spawning areas and claim small territories, where they engage in mock battles with one another. During these skirmishes the males' coloring becomes extremely intense. When the females are ripe, they move onto the spawning beds, where they are approached by the males. During the ensuing short courtships the females select their mates. Accompanied by a male, a female will wriggle into the gravel bottom so that the lower parts of her body are buried. The male then assumes a position on her back with his caudal fin next to hers. She releases the eggs in short bursts during which less than 10 eggs are released and fertilized. She then moves upstream and repeats the process of burying herself and the spawning continues in this manner until she is spent. She may continue to mate with the same male or select other mates during the spawning period. The average female will produce about 800 eggs, which are adhesive after fertilization and lodge in the crevices of the spawning gravel, where they remain until hatching.

The young grow rapidly during their first summer. By September many have reached ⅔ of their total adult length. The average life span of Rainbow Darters is about 3 years.

The diet consists mainly of small immature aquatic insects. Large adults will eat larger aquatic insects, snails, and small crayfish.

Local Names: Banded Darter, Blue Darter

102. Iowa Darter

Etheostoma exile

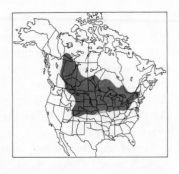

Field Marks:
- Body is slender, elongated.
- Scales on the body are medium-sized. The scales on the cheeks and gill covers are small.
- Lateral line is incomplete.
- Head is medium-sized, with a pointed snout and large eyes.
- Mouth is medium-sized, with the snout slightly overhanging the lower jaw.
- Pectoral fins are large, fan-shaped.
- Anal fin has 1 or 2 (usually 2) spines.
- First dorsal fin has 8–10 spines.
- Caudal fin is relatively small and is rounded on its trailing edge.

Adult Size: 1.5–2 in. (3.75–5.08 cm) average

Color: Females and nonbreeding males are olive or olive-brown on the back and upper sides. The lower sides are pale tan and the belly is cream. There are 8 dark brown saddle-shaped spots on the upper back and 10–14 brown or dusky vertical bands along the sides. There is a distinct dark vertical band extending downward from the eye. Ventral and anal fins are clear and unmarked. The other fins are clear and marked with dark brown lines.

Breeding males are intensely colored in shades of red, orange, yellow, and blue.

Habitats: Usually found near dense vegetation, the Iowa Darter occurs only in clear lakes and large slow-moving rivers.

Natural History: In the southern part of its range the Iowa Darter begins spawning in April and may begin as late as July in the north. The eggs are released and fertilized over a clean sand bottom, on plants rooted in sand, or on exposed tree roots beneath undercut banks. At the beginning of the spawning season, males move into shallow water and claim and defend small territories. The females remain in deep water until they are fully ripe and then move onto the spawning grounds, where they are approached by the males. Spawning takes place with the male positioned on the female's back, his anal and caudal fins beside hers. The eggs are released 3–7 at a time and are simultaneously fertilized by the male. Females may move from one mate to another during spawning and may lay as many as 600–2000 eggs. The incubation period is about 10 days at a water temperature of 55°–60°F (12.88°–15.56°C). The average life span of the Iowa Darter is 3 years. Females are usually larger than males of the same age.

The young feed mainly on plankton and very small immature aquatic insects that continue to be a staple throughout the darter's life. Mature fish eat larger insects, small crustaceans, and snails.

Local Names: Weed Darter, Yellowbelly

103. Fantail Darter

Etheostoma flabellare

Field Marks:
- Body is elongated but stout, tapering little from head to tail.
- Body scales are medium-sized. There are no scales on the head.
- Lateral line is incomplete, usually ending under the soft dorsal fin.
- Head is large, with a pointed snout and large eyes set very high on the sides of the head.
- Mouth is small with the jaws about equal in length.
- Pectoral fins are very broad, fan-shaped.
- Anal fin has 2 spines.
- First dorsal fin has 7 or 8 strong spines.
- Caudal fin is very broad, nearly square.

Adult Size: 1.5–2 in. (3.75–5.08 cm) average

Color: The back is olive-green to olive-brown. The sides below the lateral line are light tan. The belly is light tan or cream. There are 9–11 dark (usually black), vertical bars along the sides, and a dark triangle extends forward from the bottom of the eye. The base color of the dorsal and caudal fins is light olive. The second dorsal and caudal fins are marked with rows of small dark spots, usually brown or gray. The pectoral, ventral, and anal fins are very pale tan. The anal fin is marked with a few small brown spots.

Habitats: Fantail Darters can be found in the slow or moderate currents of clear streams and small rivers.

Natural History: Fantail Darters begin spawning in March or early April and continue well into June. In preparation for spawning, the males develop a series of fleshy nodes on the tips of the spines of the first dorsal fin. The males claim small territories on the downstream sides of rocks in shallow riffles. By moving about in the shallow groove between the underside of the rock and the stream bottom the male clears a small space on the rock of algae and debris. This clearing serves as the nest site. When a female is ripe, she enters the nesting areas and is approached by a male. The courtship involves the male's leading his mate to the nest and swimming close to her in a series of circles and figure 8s. Once at the rock, the female turns upside-down while the male remains upright. Then the pair swims head-to-tail, executing synchronous circles and figure 8s under the rock. When she is ready to spawn the female presses very close to the underside of the rock, vibrates slightly, and releases a single egg. The male then turns upside-down and fertilizes the egg. This process continues until the female is spent, at which time the male will usually drive her from the nest. A succession of females will spawn in each nest, each adding her eggs to the growing clutch. At the completion of spawning a nest may hold as few as 10 or as many as 500 eggs. After spawning, the male remains on the nest, protecting the eggs and keeping them clean by gently rubbing his dorsal nodes against them. At a water temperature of 70°F (21.11°C) the eggs hatch in about 20 days.

Both young and adult Fantail Darters feed mainly on small immature aquatic insects.

104. Swamp Darter

Etheostoma fusiforme

Field Marks:
- Body is slender, elongated, round in cross section.
- Scales on the body are medium-sized; on the cheeks and gill covers scales are small.
- Lateral line is incomplete.
- Head is large, with a squared snout and large eyes.
- Mouth is small, with the upper jaw slightly overhanging the lower jaw.
- Ventral fins are very long, pointed.
- Anal fin has 2 spines.
- First dorsal fin has 10 weak spines.
- Caudal fin is small, nearly square.

Adult Size: 1.5–2 in. (3.75–5.08 cm) average

Color: The back and upper sides are ochre or straw-colored. The lower sides are pale tan or light straw. The belly is pale tan or cream. The upper sides and back are marked with a sprinkling of small, irregularly shaped, dark brown or dusky spots. On the sides the spots are concentrated and form a vague broad lateral band. A dark brown or dusky band extends around the head at the level of the eyes; vague dark stripes extend down from this band. The pectoral and ventral fins are nearly transparent and unmarked. The other fins are nearly transparent and are marked with very small, dark brown spots.

Habitats: The Swamp Darter is found in slow-moving streams, ponds, and swamps, where it shelters in or near dense vegetation.

Natural History: The Swamp Darter may spawn as early as March in the southern part of its range and as late as May or June in the north. It prepares no nest and simply deposits and fertilizes the eggs on the vegetation in which it lives.

Swamp Darters feed on small immature aquatic insects and crustaceans which live with them in dense vegetation.

105. Least Darter

Etheostoma microperca

Field Marks:
- Body is elongated, delicate, slightly compressed laterally.
- Scales on the body are medium-sized. There are no scales on the head or on the breast.
- Lateral line is absent.
- Head is medium-sized, with a pointed snout and medium-sized eyes.
- Mouth is small.
- Anal fin has 1 very short and 1 long spine.
- Dorsal fins are separated but very close together.
- First dorsal fin has 6 or 7 spines.
- Caudal fin is fan-shaped, rounded on its trailing edge.

Adult Size: 1–1.5 in. (2.54–3.75 cm) average

Color: The back and upper sides are dark ochre over pale tan or light ochre on the lower sides and belly. There are several vague dark brown or dusky spots on the upper back and 8–10 squarish spots along the middle of the sides. A dark brown stripe extends downward from the eye. The pectoral, ventral, and anal fins are nearly transparent, very pale buff, and are unmarked. The other fins have the same ground color but are marked with several rows of brown spots.

Habitats: Least Darters are usually found near dense vegetation in clear lakes and rivers and in large slow-moving streams.

Natural History: The Least Darter's spawning season begins in May or June when the males move into weedy shallow water near shore. The females follow the males into the weeds, where each is met by several rival males. Brief skirmishes between the males will prove the dominance of one of them and he will accompany the female while she selects a plant on which to lay her eggs. The female assumes a vertical position along the stem of the chosen plant while the male rests on her back with his pectorals holding her ahead of her dorsal fins. The pair vibrates as 1 egg is released and fertilized and then move on to another plant and repeat the process. After several spawnings with one mate, the female returns to deeper water to rest, in preparation for another spawning with another mate. The typical female will produce 500 to 1000 eggs. The incubation period is about 6 days at a water temperature of 65°–70°F (18.48°–21.11°C).

Little is known of the diet of the Least Darter but it is assumed that it feeds mainly on tiny aquatic crustaceans.

106. Johnny Darter
Etheostoma nigrum

Field Marks:
· Body is elongated, moderately robust, nearly round in cross section.
· Scales on the body are medium-sized; on the gill covers, small. There are no scales on the cheeks or the breast.
· Lateral line is complete.
· Head is medium-sized with a blunt snout. Eyes are medium-sized.
· Mouth is small with jaws of nearly equal length.
· Pectoral and ventral fins are large. The ventrals are pointed.
· Anal fin has 1 spine.
· First dorsal has 8, 9, or 10 spines, usually 9.
· Caudal fin is long, square along its trailing edge.

Adult Size: 2–2.5 in. (5.08–6.25 cm) average

Color: The back is light brown. The sides are pale tan. The belly is very pale tan or cream. There are 6 brown saddle-shaped marks on the upper back. The sides are marked with 7–12 dusky M- or W-shaped spots along the lateral line. The pectoral, ventral, and anal fins are clear and unmarked. The dorsal and caudal fins are marked with small brown spots.

Habitats: The Johnny Darter is usually found near vegetation in shallow water in clear lakes, ponds, and slow-moving rivers. It may also be found in shallow gravel-bottomed riffles in tributary streams.

Natural History: The Johnny Darter begins its spawning season in late April, usually continuing through June. The males are first to move into shallow water, where each selects a rock or other structure under which he prepares the nest by swimming upside-down under the chosen object, cleaning it with his fins. When the ripe females enter the spawning grounds, each is met by a male as she approaches his territory. The male then returns to his nest and courts the female by swimming upside-down under the overhanging structure. After the courtship, the female joins the male, turning herself upside-down and swimming by his side. The eggs are released and fertilized one at a time as the fish continue to swim close together. Both fish usually continue spawning with each other until the female is spent. Spawning usually is completed in several periods separated by short rests.

After spawning, the male remains at the nest guarding the eggs and fanning them to remove silt or debris. During this guardianship he will eat any eggs that are diseased. The incubation period is 5–8 days at a water temperature of 70°–75°F (21.11°–24.08°C). The male guards the fry for a short time.

Johnny Darters feed mainly on small aquatic crustaceans and insects.

107. Tessellated Darter

Etheostoma olmstedi

Field Marks:
· Body is elongated, slender, nearly round in cross section.
· Scales on the body and gill covers are medium-sized. Cheeks and breast are not scaled.
· Lateral line is complete.
· Head is medium-sized, with medium-sized eyes.
· Mouth is small, with jaws of about equal length.
· Pectoral and ventral fins are large.

- Anal fin has 2 spines.
- First dorsal fin has 10 spines.
- Caudal fin is square, straight along its trailing edge.

Adult Size: 2–3 in. (5.08–7.62 cm) average.

Color: The back and upper sides are light straw-colored. The sides are pale tan or pale ochre. The belly is creamy tan. There are 6 or 7 brown saddle-shaped marks along the upper back; below these is a scattering of small, dark, V-shaped spots. Along the lateral line is a series of dusky blotches and X- or W-shaped spots. The cheeks and gill covers are marked with dusky mottling and a dusky stripe extends downward from the eye. Fins are clear or extremely light tan. All but the ventrals are marked with very small brown spots.

Habitats: The Tessellated Darter will be found in the quiet parts of small streams, rivers, or lakes over sand or mud bottom.

Natural History: The natural history of this species is nearly identical to that of the Johnny Darter (p. 174).

Local Names: Eastern Johnny Darter

108. Yellow Perch
Perca flavescens

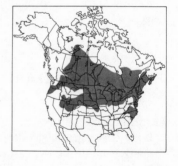

Field Marks:
- Body is elongated, robust, oval in cross section.
- Scales on the body are medium-sized. Scales on the cheeks and lower parts of the gill covers are small. Upper parts of the gill covers are not scaled.
- Lateral line is complete.
- Head is medium-sized, with a pointed snout and medium-sized eyes.
- Mouth is medium-sized, with the upper jaw ending at about the midpoint of the eye.

- Anal fin has 2 spines.
- First dorsal fin has 13–15 spines; the second dorsal fin has 1 or 2 spines. The dorsal fins are set close together but are separated.
- Caudal fin is slightly forked with triangular lobes.

Adult Size: 4–10 in. (10.16–25.4 cm) average

Color: The dark brown of the upper back extends down the sides in a series of distinct vertical bars. The dorsal and caudal fins are dark yellow or ochre. The anal, ventral, and pectoral fins are orange or yellow-orange.

Habitats: Yellow Perch flourish in a broad range of habitats, including warm or cold lakes, ponds, rivers, and streams of any size. They are tolerant of high levels of acidity and turbidity.

Natural History: Yellow Perch begin spawning in May or June. Most spawning activity takes place after dark. No nest is prepared. The spawning act involves 1 or 2 males following a ripe female as she swims over dense vegetation releasing an accordian-shaped gelatinous mass of eggs, which are fertilized by the attendant males. The egg masses swell considerably on contact with the water and become adhesive when they are fertilized. The mating fish continue swimming, allowing the eggs to settle into the vegetation. The egg masses may be 6–7 ft. (1.83–2.13 m) long and may contain 75,000 to 80,000 eggs.

The fry are slow swimmers and congregate in dense schools. At this stage of their lives, many are consumed by predators. As few as one in every 5000 Yellow Perch survives the first year of life. Most populations of Yellow Perch are relatively slow growing, with males reaching sexual maturity at 3 years and females at 4.

The fry begin life feeding on plankton and then graduate to a variety of small aquatic invertebrates, including insects and crustaceans. The adults feed heavily on insects, small crayfish, and small fishes.

Local Names: American Perch, Lake Perch, Tiger Trout

109. Logperch

Percina caprodes

Field Marks:
- Body is elongated, slender, round in cross section.
- Scales on the body are medium-sized; on the cheeks and gill covers, they are slightly smaller. The breast is not scaled.
- Lateral line is complete.
- Head is medium-sized, with a pointed snout and large eyes set high on the sides of the head.
- Mouth is small, inferior, overhung by the snout.
- Anal fin has 2 spines (rarely 1) and is longer-based than in most of the other small perches.
- Dorsal fins are separated by a very short space. The first dorsal fin has 13, 14, or 15 (usually 14) strong spines.
- Caudal fin is broad, slightly concave on its trailing edge.

Adult Size: 3–4 in. (7.62–10.16 cm) average

Color: The back is dark brown or olive. The sides and belly are dark yellow or ochre. The sides are marked with indistinct dark brown or black vertical bars. The fins are pale yellow or ochre, with dark bars on the dorsal and caudal. The other fins are unmarked.

Habitats: The Logperch is most often encountered in clear water of lakes, rivers, and creeks, usually over sand or sand/gravel bottoms.

Natural History: The Logperch's spawning season begins in June when the males gather in shallow water near shore, forming large, compact schools. The females remain in deeper water until they are fully ripe. When a female is ready to spawn she will swim into and through the school of males and one or two of them will follow her as she seeks a spawning site. She stops swimming over a patch of clean sand and settles to the bottom as the male alights on her back, holding on with his ventral fins and settling his caudal fin beside hers. Both fish vibrate very rapidly, raising a cloud of fine sand and digging themselves a little way into the bottom. The eggs and sperm are released in short bursts containing 10–20 eggs. After each spawning the female returns to deeper water and the male rejoins the school. A female will usually return to the school of males several times and mate with a different partner each time until she is spent.

As soon as the eggs are fertilized, nonbreeding males rush to the nest sites and devour any eggs that have not been buried by the activity of the mating pairs.

Little is known of the early life of the Logperch. Adults feed mainly on immature aquatic insects, especially midges; most of these are taken from the undersides of small stones, which the fish roll over with their pointed snouts.

Local Names: Manitou Darter, Zebra Fish

110. Channel Darter

Percina copelandi

Field Marks:
- Body is slender, elongated.
- Scales on the body are medium-sized; on the gill covers they are smaller. The cheeks and breast are not scaled.
- Lateral line is complete.
- Head is medium-sized, with a squarish snout and large eyes set high on the sides of the head.
- Pectoral fins are long.
- Anal fin has 2 spines.
- First dorsal fin has 11 spines.
- Caudal fin is long, straight or slightly concave on its trailing edge.

Adult Size: 1–2 in. (2.54–5.08 cm) average

Color: The back is light brown and marked with a series of small, dark brown saddle-shaped patches. The sides and belly are pale ochre with a silvery sheen. There is a series of brown or dusky spots along the middle of the sides. Three dark stripes radiate downward from the eye and one across the gill cover. The fins are all very pale sand color and are unmarked.

Habitats: Inhabiting lakes, large ponds, and large, slow-moving rivers and streams, the Channel Darter is usually found over sand or gravel bottoms near shore.

Natural History: When the water has reached a temperature of about 70°F (21.11°C), the Channel Darter begins its spawning season, which reaches its peak in July. Spawning takes place in shallow, gently flowing water. The males claim territories that include a fine gravel bottom and at least one large rock. The territories are often shared and one (or more) male will position himself behind a rock as spawning begins. Mating takes place when ripe females begin moving into the spawning territories and each male prods a mate into a position behind a rock. The male assumes a position on the female's back, grasping her back just ahead of her dorsal fin with his ventral fins and resting his caudal fin on the bottom besides hers. The pair vibrates rapidly together, which opens a shallow depression into which the eggs fall as they are fertilized. Only a few eggs are released at a time, although a female may produce 350–400 eggs. The female will often spawn with several mates before she is spent.

The Channel Darter takes most of its food from the bottom. Immature aquatic insects make up the bulk of its diet.

Local Names: Copeland's Darter

111. Blackside Darter
Percina maculata

Field Marks:
- Body is elongated, moderately robust.
- Scales on the body are medium-sized and are slightly smaller on the cheeks and gill covers.
- Lateral line is complete.

- Head is medium-sized, with a pointed snout and large eyes.
- Pectoral and ventral fins are large.
- Anal fin has 2 spines.
- First dorsal fin has 13 (sometimes 14 or 15) spines.
- Caudal fin is broad and may be straight or slightly indented along its trailing edge.

Adult Size: 1.5–2 in. (3.75–5.08 cm) average

Color: The upper back is dark ochre or dark brown. The sides are light tan or light ochre. The belly is pale tan. The back and upper sides are marked with irregular dark brown patches. There is a series of dark brown patches along the lateral line. A small dark wedge extends down from the eye, and irregular mottling is seen behind the eye on the cheek and gill cover. The first dorsal fin is flushed with dark brown or black. The second dorsal, caudal, and anal fins are marked with rows of brown spots on a pale sand-colored ground. The pectoral and ventral fins are a nearly transparent sand color and are unmarked.

Habitats: The Blackside Darter is most often encountered in clear streams with clean gravel bottoms, but it also occurs in turbid streams.

Natural History: In most of its range, the Blackside Darter's spawning activity peaks in the month of May, although it begins when the water temperature has reached 60°–65°F (15.56°–18.48°C). As the spawning season commences, males move upstream and congregate in gravel-bottomed pools, followed by the females, which take up a position in quiet water slightly downstream. When a female is fully ripe she moves into the school of males and is followed by several suitors that pursue her while she searches for a suitable nesting site, usually a shallow depression in the gravel, into which she settles. The dominant male of the pursuing group will immediately settle onto the female's back, supporting himself with his ventral fins and resting his caudal fin on the bottom next to hers. The female wriggles herself slightly into the bottom and the eggs are released and fertilized as the pair vibrates together. After a short rest period the female spawns with a different mate. The eggs are not guarded and require about 6 days for complete incubation.

Blackside Darters feed mainly on immature aquatic insects that they take from the bottom. They also occasionally take small fishes.

112. River Darter

Percina shumardi

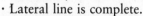

Field Marks:

· Body is elongated, moderately robust.
· Scales on the body are medium-sized, smaller on the cheeks and gill covers. The breast is not usually scaled.
· Lateral line is complete.
· Head is medium-sized, with a pointed snout and large eyes set high on its sides.
· Anal fin has 2 spines.
· First dorsal fin has 9, 10, or 11 (usually 10) spines.
· Caudal fin is long, squared or slightly concave along its trailing edge.

Adult Size: 1.5–2 in. (3.75–5.08 cm) average

Color: The back is medium brown. The sides are pale tan or ochre, and the belly is cream. The upper back is marked with small dark blotches, and there is a series of indistinct diamond-shaped spots along the lateral line. A dark wedge extends downward from the eye. The first dorsal fin has a dark spot on its leading edge and another on its trailing edge. The second dorsal fin has a row of dusky or brown spots on its membranes. The caudal fin has 3 rows of brown or dusky spots on its membranes. The other fins are unmarked. The ground color of all the fins is pale tan or pale ochre.

Habitats: The River Darter will be found in large clear rivers in areas with moderate current over gravel or boulder bottoms.

Natural History: The natural history of the River Darter is nearly identical to those of the Blackside (pp. 180–181) and Channel (pp. 179–180) Darters.

■ SCULPINS

Order Perciformes
Family Cottidae

Field Marks:

· Body is heavy near the head, slender and slightly compressed laterally near the tail.
· There are no scales on the body or head. There are usually small prickly protrusions in small patches behind the pectoral fins, along the lateral line, or over most of the body.
· Head is large and flattened and may have spines on the gill covers. Eyes are medium-sized to large and are set high on the sides of the head.
· Mouth is large.
· Eight fins: very large paired pectorals set very close behind the gill covers; paired ventrals that originate under the base of the pectorals and usually contain 1 spine; a single long-based anal; a fanlike caudal that may be squared or rounded; 2 dorsals that are separate but very close together, the first is spiny, the second is soft-rayed. With the exception of those in the caudal fin, the fin rays are not branched and look like spines at first glance.

Habitats: Most members of this family live in salt water. Those that occur in fresh water are found in clear, cool rivers, lakes, and streams. All sculpins are bottom dwellers and generally prefer a rocky or gravel bottom, free of vegetation.

Natural History: The sculpins are spring spawners that undertake short migrations into small clear streams to mate. Their spawning behavior is similar to that of the members of the perch family: The males prepare the nest and then entice the females to join them; the adhesive eggs are deposited on the underside of a large rock that serves as the nest. Usually several females spawn in each nest. The male remains on the nest after spawning to fan and guard the eggs. The fry are not immediately bottom dwellers but spend some time at midwater levels, dropping to the bottom as they grow.

The diet of the sculpins varies with the habitat and the time of year, but the principal foods are immature aquatic insects, mollusks, and crustaceans.

Collecting: Minnow traps or seines are the best tools for collecting sculpins. A seine will work best if it is staked or held in a stream while someone wades downstream toward it, overturning stones along the way.

Handling: The sculpins are cold-water species and will not thrive in temperatures over 75°F (24.08°C). They should be kept in tanks with gravel or rock bottoms and ample places to hide. Plants are not necessary. The most convenient foods for captive sculpins are bottom-dwelling aquatic worms purchased from an aquarium dealer. You may also offer them immature aquatic insects that you collect yourself. They can be trained to accept dried foods that will sink to the bottom; foods in pellet form are best for this purpose.

113. Slimy Sculpin

Cottus cognatus

Field Marks:
· Body is elongated, moderately robust near the head.
· There are small patches of prickles behind the pectoral fins.
· Lateral line is incomplete.
· Pectoral fins are very large, fanlike.
· First dorsal fin has 7 weak spines.

Adult Size: 2.5–3 in. (6.25–7.62 cm) average

Color: The ground color is light tan or straw color. The belly and chin are cream or creamy white. The sides, back, and head are marked with irregular mottlings of black or dark brown. Fins are very pale gray or tan. The ventral and anal fins are unmarked. The other fins are motled with brown and ochre.

Habitats: The Slimy Sculpin is most often found in deep cool lakes and fast-moving, gravel-bottomed rivers and streams.

Natural History: Spawning usually occurs in May at water temperatures ranging from 45° to 50°F (7.2°–10°C). The male selects a large rock in still or moving water to serve as the nest site. He then clears an area on the underside of the rock and the stream bottom below it by rubbing against the rock and sweeping the bottom with his fins. When the nest is finished he entices a mate to join him by erecting his fins and displaying his colors (which are intensified during the breeding season). He continues this display for a short time after the female has entered the nest area. When she is ready to spawn, she turns upside-down and moves under the overhanging surface of the rock where she deposits her eggs, which are immediately fertilized by the male. The fertilized eggs are adhesive. The male then drives the female from the nest and attracts another to mate with him. Several females are usually attracted to each nest. After spawning, the male assumes guardianship of the eggs and fry; this may continue even after the fry are capable of foraging on their own.

Slimy Sculpins feed on immature aquatic insects, especially stoneflies. They also consume aquatic mollusks and crustaceans.

Local Names: Muddler Minnow, Slimy Muddler, Miller's Thumb, Northern Sculpin, Rock Cusk, Brook Cusk

APPENDIX

GLOSSARY

BIBLIOGRAPHY

INDEX

Appendix: Aquatic Plants

Here is a brief list of native aquatic plants. This list is only a starting point to give a rough idea of the various types of plants that might be kept in an aquarium.

- **MOSSES AND LIVERWORTS**

  Ricciocarpus natans: Found in large masses floating on the surface of ponds and sluggish streams. Range: throughout the U.S. and southern Canada.

  Riccia fluitans: Widely used in aquariums; found floating on the surface of ponds and slow-moving streams, also at the water's edge. Range: throughout the U.S., especially in the south.

  Water moss *Fontinalis antipyretica:* Found attached to rocks in cool, clear, flowing waters. Range: northern U.S. and Canada.

- **WATER FERNS**

  Azolla caroliniana: Usually found floating on still waters (sometimes occurs along the shore in mud). Range: southeastern U.S.

- **PONDWEEDS**

  Leafy pondweed *Potamageton foliosus:* Occurs in clear streams and ponds where it will be rooted in the bottom. Range: northern North America.

  Variable pondweed *Potamageton gramineus:* Found rooted in still and slow-moving waters. Range: throughout temperate North America.

■ NAIADS

Spiny naiad *Najas marina:* Thrives in shallow, quiet, fresh and brackish waters. Range: widely distributed throughout North America; may be locally abundant or rare.

Southern naiad *Najas microdon:* Grows submerged in ponds and slow-moving streams. Range: southern U.S.

■ DUCKWEEDS

Great duckweed *Spirodella polyrhiza:* Floats in large masses on the surface of quiet streams and ponds. Range: throughout North America.

■ PICKERELWEEDS

Mud plantain *Heteranthera dubia:* Grows submerged in shallow ponds and slow-moving streams. Range: This and subspecies found throughout U.S.

Water hyacinth *Eichornia crassipes:* Floating or rooted in quiet water. Range: southern U.S.

■ WATERWEEDS

Elodea canadensis: Introduced from Europe to quiet U.S. waters where it now thrives. Range: northern North America.

■ WATER LILIES

Fanwort *Cabomba caroliniana:* Most common in shallow ponds, but also in slow streams. Range: southeastern U.S.

Watershield *Brasenia schreberi:* Ranges widely in ponds and slow-moving streams. Range: eastern and southern U.S.

■ OTHER FAMILIES

Hornwort *Ceratophyllum demersum:* Occurs in a wide range of still waters. Range: worldwide.

Glossary

Adipose fin: A small, fleshy fin with no rays or spines, originating on the midline of the back between the rear edge of the dorsal fin and the front of the caudal fin.

Aggregation: A small, dense school of fish, usually of one species.

Anadromous: Fishes that are born in fresh water, enter the sea to feed, and return to fresh water to spawn.

Anal fin: A spiny or soft-rayed fin situated on the midline of the belly behind the anus.

Anterior: Referring to the forward part of an animal's body.

Barbel: A long thin or short triangular fleshy projection found around the mouths of some species of fish.

Belly: The area of the underside of a fish's body from just behind the pectoral fins to the origin of the anal fin.

Brackish: Containing varying amounts of both fresh and salt water.

Breast: The area on the underside of a fish's body from a point between the lower parts of the gill covers to a point just below the pectoral fins.

Catadromous: A fish that is born in salt water, spends most of its time in fresh water, and returns to salt water to spawn.

Caudal fin: The tail fin of a fish.

Caudal peduncle: The area of a fish's body between the anal and caudal fins.

Cheek: The bony surface on a fish's head surrounding the lower parts of the eyes.

Crosshatch: A series of parallel lines that overlap one another to form diamond or rectangular shapes at their junctions.

Crustacean: Any one of a large group of mainly aquatic invertebrates that include lobsters and shrimps.

Ctenoid scales: Fish scales with serrations along their trailing edges.

Cycloid scales: Fish scales with smooth trailing edges.

Daphnia: A species of water flea.

Detritus: Small fragments of decomposing plant and animal matter found on the substrate of a body of water.

Dorsal: Pertaining to the back or top of an animal's body.

Dorsal fin: A fin or fins located along the midline of a fish's back, usually near the midpoint. The fin(s) may be spiny or soft-rayed.

Earflap: A thin flexible flap extending back from the rear edge of the gill covers; well developed in several species of sunfish.

Elongated: Thin, extended, having little depth.

Exotic: An imported or otherwise nonnative species of fish.

Field mark: An obvious physical characteristic that may help distinguish one species from another.

Fin ray: A thin, flexible, usually branched, rodlike structure that supports a fish's fins.

Fork length: The length of a fish's body measured from its snout to the center of the rear edge of the caudal fin.

Fry: Newly hatched fish.

Ganoid: Characteristic of certain fishes having armorlike scales of bony, enamel-covered plates, or simply a ganoid fish.

Gill: The organ with which a fish extracts oxygen from water.

Gill cover: The thin bony plate on the sides of a fish's head that protect the delicate structures of the gills.

Gill raker: A body arch whose rear edge is fringed with a dense series of tiny blood-filled filaments that absorb oxygen from the water passing over them.

Ground color: The basic color over which are markings of various patterns.

Habitat: The specific type of water in which a fish lives.

Headwater: The source of a watershed, especially a river system.

Hybrid: The offspring of parents each of which is of a different species or subspecies.

Ichthyologist: A scientist trained in the study of fishes.

Incubation: The period of time after spawning during which fertilized eggs develop to the point of hatching.

Inferior mouth: A mouth placed on the underside of the head.

Interbreeding: The mating of animals of different species or subspecies.

Invertebrate: An animal that has no backbone.

Keel: A thin hardened protrusion found on the bodies of some species of fish, usually on the midline of the belly or the lateral line.

Lateral: Situated along the side.

Lateral line: A series of small sensory pores connected to a series of scales extending along the side of a fish.

Lateral stripe: A narrow swath of color running horizontally along the side of a fish.

Laterally compressed: Flattened from side to side.

Lobe: The top or bottom half of a fish's caudal fin.

Maxillary: The upper jawbone in fishes.

Membrane: The thin, pliable part of a fish's fin that is supported by rays or spines.

Milt: A male fish's reproductive secretion that contains the sperm.

Mollusk: Any species included in a large group of invertebrate animals, such as snails or clams.

Mucus: The thin slimy coating on a fish's body.

Native: A species that is natural (not introduced) to a particular range.

Nuptial tubercles: Small fleshy nodes that develop on the heads and fins of certain species of fish (especially members of the minnow family) during the breeding season.

Offspring: The progeny (young) of animals and plants.

Operculum: The gill cover.

Ovipositor: An extension of a female fish's genitals used to deposit her eggs in a specific location.

Parr: The young of species of the salmonid group of fishes (trout and salmon).

Parr mark: A dark patch of color in a series along the sides of young salmon and trout.

Peduncle: A stalklike group of fibers connecting two parts of anatomy.

Pelagic: Living in large areas of open water.

Plankton: Plant and animal microorganisms that live suspended in water, either floating or weakly swimming.

Posterior: Referring to the rearward parts of the body.

Protrusible: Capable of being extended.

Redd: A fish nest.

Riffle: A shallow, usually gravel-bottomed, portion of a river or stream where the surface is constantly being broken by rocks and stones.

Ripe: In fishes—ready to spawn with fully developed eggs or milt.

School: A gathering of many individuals.

Scute: A bony protuberance.

Sea-run: Anadromous—having spent time in salt water.

Seine: A long, rectangular net used for catching fishes.

Sexual maturity: That time at which an animal is physically capable of reproducing.

Silt: Very fine particles of sand or clay that settle out of water and accumulate on the bottom.

Snout: The extreme forward point of a fish's head.

Soft-rayed: A fin without spines.

Spawn: In fishes—the act of mating and reproduction.

Spent: After the spawning act, a fish whose eggs or milt have all been released.

Spine: A stiff, unbranched fin ray.

Standard length: The length of a fish's body measured from the tip of the snout to the rear edge of the caudal peduncle.

Substrate: The bottom of a body of water.

Subterminal mouth: A fish's mouth that is slightly overhung by the snout.

Terminal mouth: A mouth in which the tips of the upper and lower jaws form the extreme forward point of the head.

Terrestrial: Living on the land.

Total length: The length of a fish measured from the tip of the snout to the rear edges of the caudal fin lobes.

Turbid: Water that is roiled and contains a high percentage of particulate matter.

Vertebrate: An animal with a backbone.

Ventral: Pertaining to the underside of the body, opposite the dorsal.

Ventral fin(s): The (usually) paired fins found (in most cases) near the midpoint of a fish's belly.

Vertical bar: A strip of color that runs from dorsal to ventral on a fish's body.

Water column: The depth from surface to substrate in any portion of a body of water.

Watershed: An interconnected system of bodies of water.

Year-class: All those fish of a species that are born in one year.

Bibliography

Brünner, Gerhard. 1973. *Aquarium Plants*. Neptune City, New Jersey: T. F. H. Publications, Inc.

Everhart, W. Harry. 1958. *Fishes of Maine*. Augusta, Maine: The Maine Department of Inland Fisheries and Game.

LaMonte, Francesca. 1945. *North American Game Fishes*. Garden City, New York: Doubleday and Company.

Lee, David S., R. Gilbert Carter, Charles H. Hocutt, Robert E. Jenkins, and Jay R. Stauffer, Jr. 1980. *Atlas of North American Freshwater Fishes*. Raleigh, North Carolina: North Carolina State Museum of Natural History.

McClane, A. J., editor. 1965. *Field Guide to Freshwater Fishes of North America*. New York: Holt, Rinehart and Winston.

Migdalski, Edward C., and George S. Fichter. 1976. *The Fresh and Salt Water Fishes of the World*. New York: Alfred A. Knopf.

Moyle, Peter B. 1976. *Inland Fishes of California*. Los Angeles and London: University of California Press.

Pflieger, William L. 1975. *The Fishes of Missouri*. Missouri Department of Conservation.

Reece, Maynard, 1963. *Fish and Fishing*. New York and Des Moines: Meredith Press/Better Homes and Gardens.

Scarola, John F. 1973. *Freshwater Fishes of New Hampshire*. Concord, New Hampshire: New Hampshire Fish and Game Department, Division of Inland and Marine Fisheries.

Schiotz, Arne, and Preben Dahlstrom. 1972. *A Guide to Aquarium Fishes and Plants*. Philadelphia and New York: J. B. Lippincott Company.

Scott, W. B., and E. J. Crossman. 1973. *Freshwater Fishes of Canada*. Ottawa: Fisheries Research Board of Canada.

Simon and Schuster's Complete Guide to Freshwater and Marine Aquarium Fishes. 1976. New York: Simon and Schuster.

Stodola, Jiri. 1967. *Encyclopedia of Water Plants*. Neptune City, New Jersey: T. F. H. Publications, Inc.

Thomas, G. L., Jr. 1965. *Goldfish Pools, Waterlilies, and Tropical Fishes*. Neptune City, New Jersey: T. F. H. Publications, Inc.

Trykare, Tre, and E. Cagner. 1976. *The Lore of Sportfishing*. New York: Crown Publishers, Inc.

Wachtel, Hellmuth. 1973. *Aquarium Ecology*. Neptune City, New Jersey: T. F. H. Publications, Inc.

Index

Acantharchus pomotis, 137
Adult size, 7
Aeration, 14–15
Ambloplites rupestris, 138–39
Amia calva, 31–32
Amiidae, family, 30–32
Amiiformes, order, 30–32
Ammocrypta pellucida, 165–66
Anatomy, fish, 3–6
Angling, equipment, 11–12
Anguilla rostrata, 34
Anguillidae, family, 32–34
Anguilliformes, order, 32–34
Aphredoderidae, family, 113–14
Aphredoderus sayanus, 114
Aquariums, 16–18
  furnishing, 17–18
  indoor, 16
  outdoor, 16–17
  plants, 17
Aquatic plants, 189–90
Astyanax mexicanus, 59
Atherinidae, family, 123–24
Atheriniformes, order, 116–25
Azolla caroliniana, 189

Bass, Largemouth, 159–60
  Rock, 138–39
  Smallmouth, 157–58
  See also, Sunfishes, 135–63
  White, 133–34
  Yellow, 134–35
  See also, Temperate basses, 131–35
Bluegill, 151–52
  See also, Sunfishes, 135–63
Bowfin, 31–32
Bowfins, 30–32
Brasenia schreberi, 190
Breeding fishes in the aquarium, 21–23
Bullhead, Black, 106–7
  Brown, 109
  Yellow, 108
Bullhead catfishes, 104–12

Cabomba caroliniana, 190
Campostoma anomalum, 61
Carassius auratus, 62
Carp, Common, 65–66
  See also, Minnows, 59–97
Catfish, White, 105–6
  See also, Bullhead catfishes, 104–12
Catostomidae, family, 98–101
Catostomus catostomus, 99–100
  commersoni, 100–101

Centrarchidae, family, 135–36, 139–40
Centrarchus macropterus, 139–40
Ceratophyllum demersum, 190
Characidae, family, 58
Characins, 58–59
Chub, Creek, 94–95
  Hornyhead, 71–72
  Lake, 64–65
  See also, Minnows, 59–97
Clinostomus funduloides, 63–64
Clupeidae, family, 34–35
Clupeiformes, order, 34–35
Collecting, 8
Collecting fishes, 8–13
Color, 7
Cottidae, family, 183–84
Cottus cognatus, 184–85
Couesius plumbeus, 64–65
Crappie, Black, 2, 162–63
  White, 135–36
  See also, Sunfishes, 135–63
Culea inconstans, 127–28
Cyprinidae, family, 59–61
Cypriniformes, order, 58–103
Cyprinodonidae, family, 116–17
Cyprinus carpio, 65–66

Dace, Blacknose, 91–92
  Finescale, 87–88
  Longnose, 92–93
  Northern redbelly, 86
  Pearl, 96–97
  Rosyside, 63–64

Southern redbelly, 87
  See also, Minnows, 59–97
Darter, Blackside, 180–81
  Channel, 179–80
  Eastern sand, 165–66
  Fantail, 170–71
  Greenside, 166–67
  Iowa, 169–70
  Johnny, 174–75
  Least, 173–74
  Rainbow, 167–68
  River, 182
  Swamp, 172
  Tessellated, 175–76
  See also, Perches, 164–82
Diseases, 20–21
Dorosoma cepedianum, 35–36
  petenense, 37
Duckweed, Great, 190
Duckweeds, 190

Eel, American, 34
Eichornia crassipes, 190
Elassoma evergladei, 140–41
Elodea canadensis, 190
Enneacanthus chaetodon, 142–43
  gloriosus, 143
  obesus, 144
Equipment, 9–12
Esocidae, family, 51–57
Esox americanus, americanus, 53–54
  americanus vermiculatus, 54–55
  lucius, 55–56
  niger, 56–57

Etheostoma blennioides, 166–67
  caeruleum, 167–68
  exile, 169–70
  flabellare, 170–71
  fusiforme, 172
  microperca, 173–74
  nigrum, 174–75
  olmstedi, 175–76
Exoglossum maxillingua, 67

Fanwort, 190
Fallfish, 95–96
  See also, Minnows, 59–97
Feeding fishes, 18–19
Ferns, water, 189
Field marks, 6
Filtration, 14–15
Fins, 3
Fish:
  anatomy, 3–6
    Black crappie, external
      anatomy of, 2
    Black bullhead, external
      anatomy of, 4
  breeding, 21–23
  collecting, 8–13
  diseases, 20–21
  feeding, 18–19
  handling, 12–13
  identifying, 6
  keeping, 13
  transporting, 12
  what makes a fish a fish?,
    3–5
Flier, 139–40
  See also, Sunfishes,
    135–63
Fontinalis antipyretica, 189

Freshwater eels, 32–34
  See also, Eel, American, 34
Fundulus diaphanus, 117–18
  notatus, 119

Gambusia affinis, 121
Gasterosteiformes, order,
  125–30
Gasterosteidae, family, 125–30
Gars, 27–30
  Longnose, 28–29
  Shortnose, 29–30
Goldeye, 39–40
  See also, Mooneyes, 37–39
Goldfish, 62–63
  See also, Minnows, 59–97
Gills, 5

Habitat, 7
Handling, 8
  fish, 12–13
Herrings, 34–36
Heteranthera dubia, 190
Hiodontidae, family, 37–40
Hiodon alosoides, 39–40
  tergisus, 38–39
Hornwort, 190
Hyacinth, water, 190
Hybognathus hankinsoni, 69
  nuchalis, 70
Hypentelium nigricans, 102

Identifying fishes, 6
Ictaluridae, family, 104–9
Ictalurus catus, 105–6
  melas, 106–7
  natalis, 108
  nebulosus, 109

Keeping fish, 13
Killifishes, 116–19
    Banded, 117–18
    See also, Blackstripe top-
      minnow, 119

Labidesthes sicculus, 124–25
Lateral line, 5
Leafy pondweed, 189
Lepisosteidae, family, 27–29
Lepisosteiformes, order, 27–
  29
Lepisosteus osseus, 28
  platostomus, 29
Lepomis auritus, 145–46
  cyanellus, 146–47
  gibbosus, 147–48
  gulosus, 149–50
  humilis, 150–51
  macrochirus, 151–52
  megalotis, 152–53
  microlophus, 154–55
  punctatus, 155–56
  symmetricus, 156
Livebearers, 120–22
Local names, 8

Madtom, Brindled, 112
  Tadpole, 111
  See also, Bullhead
    catfishes, 104–12
Micropterus dolomieui, 157–
  58
  salmoides, 159–60
Minnows, 59–97
  Bluntnose, 88–89
  Brassy, 69
  Cutlips, 67–68
  Fathead, 90–91

  silvery, Central, 70
  See also, Chub, 64–65,
    71–72, 94–95; Com-
    mon Carp, 65–66;
    Dace, 63–64, 86, 87–
    88, 91–93, 96–97;
    Goldfish, 62–63;
    Shiner, 72–85;
    Stoneroller, 61–62
Minnow traps, 11
Minytrema melanops, 103
Molly, Sailfin, 122–23
  See also, Livebearers,
    120–22
Mooneye, 38–39
Mooneyes, 37–39
Morone americana, 132–33
  chrysops, 133–34
  mississippiensis, 134–35
Mosquitofish, 121–22
  See also, Livebearers,
    120–22
Mosses, 189
  Water, 189
Mudminnows, 49–51
  Central, 50–51
  Eastern, 51
Mud plantain, 190

Naiads, 190
  Southern, 190
  Spiny, 190
Najas marina, 190
  microdon, 190
Natural history, 7
Nocomis biguttatus, 71
Notemigonus crysoleucas, 72
Notropis atherinoides, 73–74
  bifrenatus, 74–75

chalybaeus, 75–76
coccogenis, 76–77
cornutus, 77–78
galacturus, 78–79
hudsonius, 79–80
rubellus, 80–81
spilopterus, 81–82
stramineus, 82–83
vollucellus, 83–84
zonatus, 84–85
Noturus flavus, 110
gyrinus, 111
miurus, 112

Osteoglossiformes, order, 37–40

Perca flavacens, 176–77
Perches, 164–82
Logperch, 178–79
Yellow, 176–77
White, 132–33
See also, Temperate Basses, 131–35
Percichthyidae, family, 131–34
Percidae, family, 164–83
Perciformes, order, 131–35, 164–83
Percina caprodes, 178–79
copelandi, 179–80
maculata, 180–82
shumardi, 182
Percopsidae, family, 114–16
Percopsiformes, order, 113–14
Percopsis omiscomaycus, 115–16
Phoxinus eos, 86
Pickerel, Chain, 56–57
Grass, 54–55

Redfin, 53–54
See also, Pikes, 51–57
Pickerelweeds, 190
Pikes, 51–57
Northern, 55–56
Pimephales notatus, 88–89
promelas, 90–91
Pirate Perch, 114–15
Pirate Perches, 113–14
Plants, aquatic, 189–90
in the aquàrium, 17
Poecilia latipinna, 122–23
Poeciliidae, family, 120–23
Pomoxis annularis, 161–62
nigromaculatus, 162–63
Pondweeds, 189
Leafy, 189
Variable, 189
Potamageton foliosus, 189
gramineus, 189
Pumpkinseed, 147–48
See also, Sunfishes, 135–63
Pungitius pungitius, 129–30

Rhinichthys atratulus, 91–92
cataractae, 92–93
Riccia fluitans, 189
Ricciocarpus natans, 189

Salmoniformes, order, 40–42, 49–50, 51–53
Salmonidae, family, 40–47
Salmo aguabonita, 42–43
clarki, 44
gairdneri, 45
trutta, 46–47
Salvelinus fontinalis, 47–48
Scales, 3–5

Sculpins, 183–85
  Slimy, 184–85
Semotilus atromaculatus, 94–95
  corporalis, 95–96
  margarita, 96–97
Shad, Gizzard, 35–36
  Threadfin, 37
  See also, Herrings, 34–37
Shiner, Bleeding, 84–85
  Bridle, 74–75
  Common, 77–78
  Emerald, 73–74
  Golden, 72–73
  Ironcolor, 75–76
  Mimic, 83–84
  Rosyface, 80–81
  Sand, 82–83
  Spotfin, 81–82
  Spottail, 79–80
  Warpaint, 76–77
  Whitetail, 78–79
  See also, Minnows, 59–97
Siluriformes, order, 104–12
Silversides, 123–25
  Brook, 124–25
Southern naiad, 190
Spiny naiad, 190
Sticklebacks, 125–30
  Brook, 127–28
  Ninespine, 129–30
  Threespine, 128–29
Stonecat, 110
  See also, Bullhead
    catfishes, 104–5
Stoneroller, 61–62
  See also, Minnows, 59–97
Suckers, 98–103
  Longnose, 99–100

Northern hog, 102
Spotted, 103
White, 100–101
Sunfishes, 135–63
  Bantam, 156
  Banded, 144
  Banded pygmy, 141–42
  Blackbanded, 142–43
  Bluespotted, 143
  Everglades pygmy, 140–41
  Green, 146–47
  Longear, 152–53
  Mud, 137
  Orangespotted, 150–51
  Redbreast, 145–46
  Redear, 154–55
  Spotted, 155–56

Temperate basses, 131–34
Tetra, Mexican, 59
  See also, Characins, 58–59
Topminnow, Blackstripe, 119
  See also, Killifishes, 116–19
Transporting fish, 12
Trout-perch, 115–16
Trout-perches, 114–16
Trouts, 40–48
  Brook, 47–48
  Brown, 46–47
  Cutthroat, 44
  Golden, 42–43
  Rainbow, 45

Umbra limi, 50
  pygmaea, 51
Umbridae, family, 49–51

Variable pondweed, 189

Warmouth, 149–50
    See also, Sunfishes, 135–
        63
Water, 14–16
    temperature, 15
    quality, 15
Water ferns, 189
Water hyacinth, 190

Water lilies, 190
Water moss, 189
Watershield, 190
Waterweeds, 190
What makes a fish a fish?, 3–
    5
White bass, 133–34
White perch, 132–33
    See also, Temperate basses,
        131–34